Advances in

BOTANICAL RESEARCH

incorporating *Advances in Plant Pathology*

VOLUME 37

ANTHOCYANINS IN LEAVES

Advances in

BOTANICAL RESEARCH

incorporating *Advances in Plant Pathology*

Advances in

BOTANICAL RESEARCH

incorporating *Advances in Plant Pathology*

ANTHOCYANINS IN LEAVES

Edited by

K. S. GOULD

School of Biological Sciences,
University of Auckland, New Zealand

and

D. W. LEE

Department of Biological Sciences,
Florida International University,
Miami, USA

Series Editor

J. A. CALLOW

School of Biosciences,
University of Birmingham,
Birmingham, UK

VOLUME 37

2002

ACADEMIC PRESS

An Imprint of Elsevier Science

Amsterdam Boston London New York Oxford Paris
San Diego San Fransisco Singapore Sydney Tokyo

This book is printed on acid-free paper.

Academic Press
An imprint of Elsevier Science
84 Theobald's Road, London WCIX 8RR, UK
http://www.academicpress.com

Academic Press
An imprint of Elsevier Science
525 B Street, Suite 1900, San Diego, California 92101-4495, USA
http://www.academicpress.com

ISBN 0-12-005937-1

A catalogue record for this book is available from the British Library

Typeset by EXPO Holdings, Malaysia
Printed and bound in Great Britain by MPG Books Limited, Bodmin, Cornwall
01 02 03 04 05 06 MP 9 8 7 6 5 4 3 2 1

CONTENTS

Anthocyanins in Leaves: Distribution, Phylogeny and Development

D. W. LEE

The Final Steps in Anthocyanin Formation: A Story of Modification and Sequestration

C. WINEFIELD

Molecular Genetics and Control of Anthocyanin Expression

B. WINKEL-SHIRLEY

Differential Expression and Functional Significance of Anthocyanins in Relation to Phasic Development in *Hedera helix* L.

W. P. HACKETT

Do Anthocyanins Function as Osmoregulators in Leaf Tissues?

L. CHALKER-SCOTT

The Role of Anthocyanins for Photosynthesis of Alaskan Arctic Evergreens during Snowmelt

S. F. OBERBAUER and G. STARR

Anthocyanins in Autumn Leaf Senescence

D. W. LEE

A Unified Explanation for Anthocyanins in Leaves?

K. S. GOULD, S. O. NEILL and T. C. VOGELMANN

The colour plate section appears between pages 106 and 107

CONTRIBUTORS TO VOLUME 37

L. CHALKER-SCOTT *Division of Ecosystem Sciences, College of Forest Resources, Box 354115, University of Washington, Seattle, WA 98195, USA*

T. S. FEILD *Department of Integrative Biology, University of California, Berkeley, CA 94720-3140, USA*

K. S. GOULD *School of Biological Sciences, University of Auckland, Private Bag 92019, Auckland, New Zealand*

W. P. HACKETT *Department of Environmental Horticulture, University of California, Davis, CA 95616, USA*

N. M. HOLBROOK *Department of Organismic and Evolutionary Biology, Harvard University, Cambridge, MA 02138, USA*

D. W. LEE *Department of Biological Sciences, Florida International University, Miami, FL 33199, USA and Fairchild Tropical Garden, Miami, FL 33156, USA*

S. O. NEILL *School of Biological Sciences, University of Auckland, Private Bag 92019, Auckland, New Zealand*

S. F. OBERBAUER *Department of Biological Sciences, Florida International University, Miami, FL 33199, USA*

G. STARR *School of Forestry Resources and Conservation, University of Florida, Gainesville, FL 32611, USA and Department of Biological Sciences, Florida International University, Miami, FL 33199, USA*

G. S. TIMMINS *College of Pharmacy, University of New Mexico, Albuquerque, NM 87131, USA*

T. C. VOGELMANN *Department of Botany and Agricultural Biochemistry. The University of Vermont, Burlington, VT 05405, USA*

C. WINEFIELD *New Zealand Institute for Crop and Food Research Ltd, Private Bag 11600, Palmerston North, New Zealand and current address Biology Department, University of York, P.O. Box 373, Heslington Road, York YO10 5YW, UK*

B. WINKEL-SHIRLEY *Department of Biology, Virginia Tech, Blacksburg, VA 24061-0406, USA*

CONTENTS OF VOLUMES 27–36

Contents of Volume 27

Contents of Volume 28

Contents of Volume 29

Contents of Volume 30

Contents of Volume 31

Contents of Volume 32

Contents of Volume 33

Contents of Volume 34

Contents of Volume 35

Contents of Volume 36

PREFACE

The chapters in this volume are the product of a symposium entitled 'Why leaves turn red: the function of anthocyanins in vegetative organs', that was held at the 2001 meeting of the the Botanical Society of America, held in Albuquerque, New Mexico, USA, in August of 2001. We are indebted for the financial support of the Society, and the sponsorship of both the Phytochemistry and the Structure and Development sections. Two participants in those meetings were not able to prepare chapters for this volume, but contributed by their stimulating presentations, and by providing ideas and feedback during the meetings and during the preparation of chapters. Mark Rausher provided a broader evolutionary context for our molecular, biochemical and physiological discussions, and Virginia Walbot was a gold mine of information on all matters concerning anthocyanins. Robert Wallace, the Chairperson of the Phytochemical Section, helped to organize the symposium.

Kevin S. Gould
David W. Lee

Anthocyanins in Leaves and Other Vegetative Organs: An Introduction

DAVID W. LEE[1,2] and KEVIN S. GOULD[3]

[1]Department of Biological Sciences, Florida International University, Miami, FL 33199, USA
[2]Fairchild Tropical Garden, Miami, FL 33156, USA
[3]School of Biological Sciences, University of Auckland, Private Bag 92019, Auckland, New Zealand

ABSTRACT

Although anthocyanins are most recognized as pigments contributing to coloration in fruits and flowers, they are also present in leaves and other vegetative organs. Although their presence has long been recognized, particularly because of their contribution to autumn coloration, the phenomenon has been poorly studied and is not well understood. In this chapter we review the history of research on anthocyanins in leaves, emphasizing the flurry of research at the end of the 19th century as well as the growing body of contemporary research on the topic. We emphasize the various hypotheses of anthocyanin function that were mainly developed more than a century ago, and emphasize recent research that takes advantage of our dramatically increased understanding of whole plant physiology.

Advances in Botanical Research Vol. 37
incorporating Advances in Plant Pathology
ISBN 0-12-005937-1

I. INTRODUCTION

Anthocyanins (Fig. 1) are red, purple or blue pigments derived from the flavonoid pathway. They are synthesized in bryophytes and vascular plants, most notably among the angiosperms, but are absent from the algae. In the flowering plants, anthocyanins are most conspicuous in the coloration of flowers and fruits, although they are also to be found in the leaves, stems and roots of many species. The economic importance of anthocyanins to the fruit and cut-flower industries has driven intensive research into the molecular control of cyanic coloration (Mol *et al.*, 1996). However, despite the widespread distribution of anthocyanins among vegetative organs, and the long history of interest in anthocyanins, relatively little research has been directed towards their function in leaves. There are several reasons for this lack of research; these are best understood by reviewing the history of study of red pigmentation in plants.

II. HISTORY

A. EARLY

Interest in the red coloration of leaves and other plant parts can be traced back to the beginnings of western science. Red coloration in leaves was observed by Aristotle (Barnes, 1984) in his writings on the theory of colour, and by Theophrastus (1918) in his descriptions of different plants. During the renaissance, Nehemiah Grew (1682) observed the red pigmentation in vegetative organs and conducted simple experiments on them, described in his "The Anatomy of Plants". Various scientists observed the distribution of 'coloured cell sap' in plant organs with the light microscope, for which Marquart (1835) coined the term anthocyanin, derived from the Greek *anthos* (flower) and *kyanos* (blue). Many long-standing misconceptions on anthocyanin function were derived

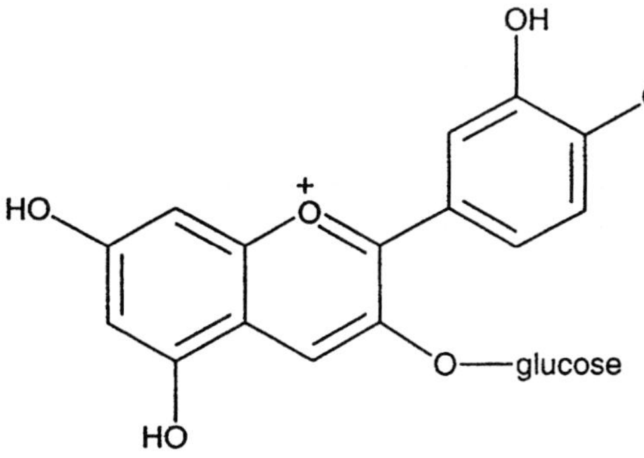

Fig. 1. The structure of cyanidin-3-glucoside, the most commonly encountered anthocyanin in leaves.

from these early observations. For instance von Mohl (1837) held that anthocyanins were derived from the breakdown of chlorophyll, based on his observations of autumn leaf coloration. Clearly, this is incorrect, yet the adjunct hypothesis, that anthocyanins are revealed during leaf senescence as chlorophylls degrade (which is also incorrect) persists in the contemporary literature. The early history of anthocyanin research was reviewed in detail by Wheldale (1916), many of whose findings are highlighted in this chapter.

B. LATE 19TH AND EARLY 20TH CENTURY

The dramatic expansion of research on anthocyanins during this period can be ascribed to three factors. First, there was the rise of organic chemistry, particularly in Germany, that made possible the determination of the chemical nature of anthocyanins and other pigments. Second was the development of a 'school' of plant research in Europe known as physiological plant anatomy. Third was the re-discovery of Mendel's laws of inheritance.

The attachment of anthocyanins to sugar molecules was first described by Lidforss in 1891 (Wheldale, 1916). Improvements in the techniques for the elucidation of chemical structures culminated in the classical studies of Willstatter and colleagues in which the formula and structure of an anthocyanin molecule was determined, and a biochemical pathway of its synthesis was proposed (Willstätter and Everest, 1912; Willstätter and Mallison, 1914).

During this time, work on anatomy, morphology and physiology in plants became fused as a new discipline: physiological plant anatomy (Haberlandt, 1914; Cittadino, 1990). Much of the inspiration for this research was the experience of European botanists in the tropics, particularly in Southeast Asia. There, they observed the presence of red pigmentation in tropical plants, particularly the rapidly expanding foliage of trees and the undersurfaces of shade plants. The striking coloration in leaves of these plants, rarely observed in temperate regions where speculation was primarily derived from observations of autumn leaf colour, stimulated much research on anthocyanin function in leaves.

Three hypotheses on the effects of anthocyanins on leaf function were proposed at this time: photoprotection of photosynthesis; photoprotection of starch hydrolysis and translocation; and temperature regulation.

The German physiologist Pringsheim (1879–1882) had shown that chlorophyll in leaves was damaged by exposures to white light of high intensity, but not to red light. This led him to suggest that anthocyanin might protect the photosynthetic machinery by screening out the most damaging wavelengths of light, thus establishing the photoprotection hypothesis. He was influenced by many earlier observations, summarized by Kerner von Marilaum (1897), that sun exposure promoted the synthesis of red pigmentation. Others noted the relationship between sugar accumulation and anthocyanin formation, and hypothesized that anthocyanins might protect against damaging effects of intense light on starch hydrolysis and sugar translocation (Pick, 1883).

Since anthocyanins absorb radiation at wavelengths different than chlorophyll, some scientists argued that these pigments might moderate leaf temperature. The most prominent among the physiological plant anatomists was Ernst Stahl. Based on his research in Java, Stahl (1896) concluded that anthocyanins could contribute to increased absorption of light by leaves, thereby elevating leaf temperatures and increasing rates of transpiration and metabolism. Stahl's hypothesis directly opposed that of Keeble (1895), who had previously suggested that in the flushing and hanging red leaves of tropical trees, high temperatures would be moderated by the red pigments. The publication of Stahl's hypothesis stirred controversy among the scientific community. In Ewart's (1897) critique, Stahl's hypothesis was denounced, and evidence presented instead in favour of the photoprotection hypothesis. Direct evidence supporting Stahl's hypothesis was provided by Smith (1909) who, using thermocouple sensors to measure leaf temperatures, reported that red leaves were up to several degrees warmer than green leaves under comparable environmental conditions.

Shortly after the turn of the century, the laws of Mendel were re-discovered. Contemporaneous discoveries in the chemistry of anthocyanins and other flavonoid pigments led to the use of these pigments as biochemical characters in genetics. The control of flower and fruit colour presented an economic incentive for research in the biochemical genetics of anthocyanins. This was well documented by Wheldale (1916, later as Onslow, 1926), who was an active participant in the research. Advances in the inheritance of anthocyanins led to greater understanding of the pathways of synthesis of these pigments, and this research continued during the first half of the 20th century.

C. THE LAST FIFTY YEARS

Dramatic advances in the understanding of flavonoids and anthocyanins followed the Second World War, particularly due to the incorporation of carbon 14 into tracer molecules to resolve pathways of synthesis. This research was particularly centred in the laboratory of Hans Grisebach at the University of Freiburg (Grisebach, 1957; Grisebach and Patschke, 1961), although others also contributed (Geismann and Swain, 1957; Watkin *et al.,* 1957). Much of this research was completed in the 1950s, and the pathway was further refined with the advent of high performance liquid chromatography in the 1980s (Stafford, 1990). During this period of chemical research, there were few studies on the functions of anthocyanins in leaves – only their roles in pollinator attraction to flowers and disperser attraction to fruits were considered.

D. WHY NOW?

The bulk of research on anthocyanins has concentrated on those plant organs of greatest economic significance: flowers and fruits. There has been, and continues to be, considerable commercial incentive to control flower colour for floriculture, and fruit colour for the horticultural industry. However, the past decade has also seen a significant increase in research on the functions of anthocyanins in leaves and other vegetative organs. There are several reasons for the long stagnant period and the recent resurgence in activity. These relate to the history of research, economic pressures, trends towards reductionist approaches in biology, and recent advances in whole-plant physiology and molecular genetics.

Early observations that anthocyanins in leaves were inducible by light led to an expansion in research into plant photobiology. The pathway of anthocyanin biosynthesis, with the induction of enzymes at different points, has became a model system in photobiology (Mancinelli, 1985; Beggs and Wellman, 1994) and plant molecular biology (Westhoff, 1998), and is a key component of research on phytochrome (Sage, 1992). Thus, it is not surprising that we have a detailed knowledge of the biosynthesis of these pigments, including the molecular genetics of their control at different points in their biosynthetic pathway. In contrast, we have relatively little knowledge of anthocyanin function(s) in plant organs, other than their potential to attract animals for pollination and seed dispersal. Moreover, the focus on important crop plants has meant neglect in research on the functions of these molecules in vegetative organs of non-crop plants. There is a parallel in the history of research on anthocyanins in that for phytochrome. The discovery of the red:far-red effects led to the elucidation of the pigment molecule, and eventually to its molecular biology. Yet, research on the function of phytochrome in nature (Smith, 1994) lagged far behind.

In the past decade, interest in the questions of anthocyanin function in vegetative organs has definitely increased. There are several reasons for this. First, there have been steady advances in whole plant physiology and physiological ecology, partly made possible by the development of sophisticated field-portable instrumentation. Foremost among these are fluorometers used in dissecting photosynthetic function by the analysis of the kinetics of chlorophyll fluorescence decay. We now have a more detailed understanding of the impact of high irradiance on photosynthetic processes, and the potential mitigating effects of anthocyanins as light filters. Second, with the realisation that global levels of UV-B radiation are rising, researchers have actively searched for UV-screening agents. The UV-absorbing properties of most flavonoids, including those of the anthocyanins, are obvious targets for research (Shirley, 1996). A third development is the growing awareness of the dietary importance of flavonoids as antioxidants (Rice-Evans and Packer, 1998). Anthocyanins in particular are potent scavengers of most reactive oxygen species. Given their documented effects in human diet, the scientific community has begun also to realise the importance of anthocyanins in the amelioration of plant defence responses to oxidative damage.

III. HYPOTHESES FOR ANTHOCYANIN FUNCTION IN LEAVES

Given the long, if somewhat fragmented, history of research into anthocyanins, it is not surprising that many different hypotheses have been proposed for the functions of these pigments in leaves. Here we summarize the chief hypotheses that have been debated over the past decade or so. Since most of these are explored in far greater depth in the following chapters, we succinctly summarize them and expand on those not covered in other chapters.

A. UV-B PROTECTION

Anthocyanins, particularly acylated forms, absorb UV-B radiation (Harborne, 1988), are induced and accumulate in plants subjected to UV-B (Kakegawa *et al.*, 1991; Klaper *et al.*, 1996) and reduce photoinhibition and DNA damage in UV-irradiated material (Takahashi *et al.*, 1991; Stapleton and Walbot, 1994; Burger and Edwards, 1996; Klaper *et al.*,

1996). Flavonoid-deficient mutants are hypersensitive to UV-B (Li *et al.*, 1993; Lois and Buchanan, 1994). However, anthocyanins are less effective UV-protectants than other flavonoids and hydroxycinnamic acid cunjugates (Woodall and Stewart, 1998) and in many leaves anthocyanins are not optimally located for UV-B screening (Lee *et al.*, 1987; Gould and Quinn, 1999; Lee and Collins, 2001) compared to other flavonoids (Burchard *et al.*, 2000).

B. LEAF WARMING

Quanta absorbed by anthocyanins may be converted into heat, serving to elevate leaf temperatures. Cyanic leaves are common in cold locations, particularly at high altitudes (McClure, 1975; Ganders *et al.*, 1980; Hoch *et al.*, 2001). Modest temperature increases were noted for cyanic versus green conifer needles (Sturgeon and Mitten, 1980), though no differences were found among tropical species (Lee *et al.*, 1979, 1987). Lee *et al.* (2002) found no differences in leaf temperatures between senescing green and red leaves of *Quercus rubra* and *Vaccinium corymbosum* in a New England forest. Anthocyanins are not required for freezing tolerance in *Arabidopsis* (Leyva *et al.*, 1995; McKown *et al.*, 1996). This hypothesis will be discussed in this volume in chapters by Chalker-Scott as well as by Starr and Oberbauer.

C. DEFENCE AGAINST HERBIVORES AND PATHOGENS

As an end-product of the flavonoid pathway, anthocyanins have often been assumed to be biologically active along with other flavonoid compounds. Insect attacks may induce the production of anthocyanins around the lesions on leaves (Costa-Arbulú *et al.*, 2001), as they are produced in response to mechanical damage and fungal attack. However, the direct evidence for such activity is not very strong. Nonetheless, anthocyanins have been hypothesized to function defensively for such putative activity, and also in their alteration of the appearances of leaves.

A wide variety of flavonoid compounds has been implicated in defensive activity against herbivores, particularly as toxins and anti-feedants against insects, and against fungal pathogens, often as phytoalexins (Rosenthal and Janzen, 1979; Harborne, 1997; Lambers *et al.*, 1998). However, such evidence is not strong. In fact anthocyanins are known to be well-tolerated in the diets of higher animals, including humans.

Anthocyanins are frequently produced around lesions from fungal pathogens (Harborne, 1976; Chalker-Scott, 1999 and this volume). This is particularly well-documented for diseases of maize, for instance by Hammaerschmidt and Nicholson (1977). However, evidence for actual inhibition of fungal growth is weak. Interestingly, strong evidence for the antifungal activity of a 3-deoxy anthocyanin has been shown for soybeans against several pathogens (Nicholson *et al.*, 1987), but not for anthocyanins.

Coley and Aide (1989) reported leaf cutting insects (that remove leaf parts and carry them to nests to cultivate fungal gardens) in feeding tests to discriminate against both young red leaves and food laced with anthocyanic leaf extracts. They speculated that such a food collection strategy would logically protect their fungal gardens against antifungal activity. However, they did not directly demonstrate such activity or compare the total flavonoid chemistry of the leaf pieces. Numerous attempts to show such antifungal activity for anthocyanins in a variety of contexts have produced negative results (Harborne and Grayer, 1988; Hoagland and Boyette, 1994).

Similar attempts to show biological effects on herbivorous insects have also largely been unsuccessful. However, Hedin and colleagues (1983) showed that cyanidin-3 glucoside, commonly produced in leaves, including cotton, inhibited the growth of larvae of the tobacco budworm, *Heliothis viridis*, an important pest of cotton and other crops. This is an exception to the general lack of results of inhibitory or toxic activity by anthocyanins.

Anthocyanins may have a more important role in reducing herbivory by altering leaf appearance. Anthocyanins and chlorophylls combine to produce brown colours in leaves, almost black if in sufficient concentrations. Of course, in the absence of chlorophylls and in the right placement in the leaf, anthocyanins produce a very bright colour at rather low concentrations, given the high extinction coefficients of anthocyanins in the wavelengths around 510–550 nm and absorbances of other ubiqutous pigments at shorter wavelengths. Stiles (1982) has argued that red leaves attract dispersal agents when associated with less conspicuous ripe fruits.

Stone (1979) hypothesized that the accumulation of anthocyanins in young fronds of certain species of palms could disguise the leaves and thus reduce rates of herbivory during development. Some rainforest understory herbs produce mature leaves with combinations of chlorophylls and anthocyanins, and the leaves are indistinguishable from the adjacent dead leaves, such as *Psychotria ulviformis* in French Guyana (unpublished research). Juniper (1993) also argued that such coloration could disguise young leaves of temperate plants during the growing season, to make them less susceptible to herbivores. Another possible camouflage of leaves is variegation, not exclusively but often through the production of anthocyanic spots. Givnish (1987, 1990) argued that such spots may make leaves a difficult search image for potential herbivores. Smith (1986) had shown in *Byttneria aculeata* (Sterculiaceae, now Malvaceae), a small understory herb of the neotropics, that variegation could reduce herbivory by leaf mining insects as well as improve the energy properties of the leaves, depending on the location of the plants. This is an important study because it is unique in backing up this hypothesis with experimental evidence. However, the source of variegation was not due to anthocyanins. Ganders *et al.* (1980) worked out the controls of inheritance of variegation due to anthocyanin spots in three coniferous forest herbs. They observed gradients in the frequency of variegation, but concluded that such spots increased leaf temperatures and did not affect herbivory. Variegation, often due to anthocyanin spots, is common among understory plants in both temperate and, particularly, tropical forests. We need more experiments and less speculation.

In the absence of chlorophylls, anthocyanins could also make leaves more apparent to potential herbivores, and warn them away. Such an appearance could warn of some negative consequence, as poor nutrition or toxicity. Coley has argued that the young leaves of many tropical trees delay greening (and the production of nutritious tissue) as a strategy to reduce herbivory during leaf expansion. Such leaves frequently produce anthocyanins and are brilliantly colored during expansion. This is a commonly observed phenomenon in tropical trees. She has speculated that such coloration may be a signal of this low palatability and deter herbivores (Kursar and Coley, 1992; Coley and Barone, 1996). In a large survey of tropical woody taxa, the majority of species with flushing red leaves produced appreciable levels of chlorophyll during development (Lee and Collins, 2001) and two well-known red-flushing species, mango and cacao, produce increasing amounts of chlorophylls during leaf expansion (Lee *et al.*, 1987). Before his untimely death, William Hamilton developed the hypothesis that red autumn coloration protected against egg-laying and future herbivory by aphids through warning. Brown and Hamilton (2001) recently reported correlation between incidence of aphid attacks and anthocyanic coloration in a survey of 262 tree species. However, their conclusions are based on surveys of the literature and not on direct observation and experimentation. Archetti (2000) has constructed a model of coevolution that demonstrates how such a phenomenon could occur.

Thus, the hypotheses of anthocyanin involvement in defence against attack by pathogens and herbivores are interesting, but not very well supported by experimental data.

D. DROUGHT RESISTANCE

Anthocyanins may confer resistance to water stresss by osmotic adjustment of the vacuolar sap (Chalker-Scott, 1999 and this volume). Cell cultures and whole-plant systems subjected to osmotic stressors often accumulate anthocyanins (Do and Cormier, 1991; Murray *et al.*, 1994; Tholakalabavi *et al.*, 1997; Mita *et al.*, 1997). Plants that are drought-tolerant commonly have cyanic leaves (Bahler *et al.*, 1991; Sherwin and Farrant, 1998).

E. LIGHT BACKSCATTERING

Abaxially located anthocyanins might enhance light capture in understory plants by reflecting photons that would otherwise penetrate the leaf back to the chlorenchyma (Lee *et al.*, 1979). The hypothesis is mechanistically difficult to understand, and has been refuted (Gould *et al.*, 1995; Neill and Gould, 1999).

F. PHOTOPROTECTION

By absorbing quanta that would otherwise be intercepted by chloroplasts, anthocyanins may reduce both the requirements for non-photochemical quenching, and the structural damage associated with chronic photoinhibition under high irradiances and low temperatures. Enhanced photosynthetic performance under strong light has been reported for red versus green leaves in several species (Gould *et al.*, 1995; Krol *et al.*, 1995; Smillie and Hetherington, 1999; Hoch *et al.*, 2001; Feild *et al.*, 2001), although no differences were found among red- and green-leafed *Coleus* varieties (Burger and Edwards, 1996). This hypothesis will be discussed in some detail in the chapter by Timmins *et al.*, in this volume.

G. ANTIOXIDANT PROTECTION

Anthocyanins may mitigate oxidative damage in leaves subjected to biotic or abiotic stress (Yamasaki, 1997; Neill *et al.*, 2002). Anthocyanins scavenge most species of reactive oxygen (Bors *et al.*, 1994). They could also reduce photooxidative stress by reducing the light flux incident on chloroplasts (Neill, 2002), and by chelating transition metals in the cell vacuole (van Acker *et al.*, 1996). This hypothesis will be discussed in more detail in the chapter in this volume by Gould *et al.*

IV. AN EVOLUTIONARY PERSPECTIVE

All of the above hypotheses predict advantages for plants as protection against the loss of photosynthetic tissue, thereby enhancing the ability of plants to assimilate carbon over the long run. However, to show that such a mechanism is evolutionarily significant it is necessary to demonstrate that such a mechanism increases the ***fitness*** of the organism, seen as the greater survival and/or reproduction of offspring with such a defensive feature, over a period of time. We have virtually no such evidence of function of anthocyanins in such an evolutionary context.

Given the appearance of anthocyanins, as part of flavonoid metabolism, with the arrival of terrestrial plants, well over 400 million years ago, it seems that there should be one or more such functions. Their current importance in attracting pollinators and dispersers would not have been possible 250 million years before the arrival of the angiosperms. It seems probably that one or more of these explanations (or perhaps others not mentioned) could have functioned in these early plants, and these pigments could have assumed other functions much later. Research providing evidence on a molecular, cellular, or integrated physiological level is an important first step. However, it is useful to speculate on the kind of research that will be necessary to put hypotheses of anthocyanin function in a stronger evolutionary perspective.

Research on the evolutionary significance of anthocyanin function will require the performance of traits of anthocyanin production in plant organs in natural field settings. Mark Rausher and collaborators (Fineblum and Rausher, 1997) have demonstrated the kinds of experiments that can yield evidence of evolutionary significance. Such research requires the understanding of the genetic control of anthocyanin production in plant organs, and then the evaluation of the field performance of such genes. This research has also revealed the importance of considering the interactive or conflicting effects of anthocyanins produced in different organs. Thus, a gene producing an anthocyanic flower may also produce purple leaves. The anthocyanic flowers may be better pollinated (and leave more progeny) while the anthocyanic leaves could also influence rates of herbivory and affect the numbers of progeny. Simms and Bucher (1996) did detect such pleiotropic effects of flower colour intensity on herbivore performance in *Ipomoea purpurea*.

Mutants for anthocyanin production in plant organs can be used in research on function (and fitness) in two ways. Careful observations within natural populations can reveal considerable variation in the production of anthocyanins in vegetative organs. Breeding experiments can determine the patterns of inheritance, and pure lines can be used for the kinds of physiological research described in this volume. Then these plants can be grown in mixed populations in semi-natural settings to determine the production of progeny and their survivorship. Secondly, mutants (particularly those ‘knocking out’ anthocyanin production at specific points in the biosynthetic pathway) can be developed using the techniques of molecular genetics to produce plants for physiological and population experiments.

In certain cases natural selection may have already 'performed' such experiments and the comparative biological approaches may provide useful evidence. For instance, the hypothesis of protective anthocyanin function during leaf senescence (see 'Anthocyanins in Autumn Leaf Senescence' in this volume) stipulates that such protected leaves will have less nitrogen in tissues at leaf fall because relatively more has been resorbed by the parent plant. A comparative study could focus on a taxon of closely related species, with annual and perennial growth habits. We would expect that the annuals would not produce anthocyanins during leaf senescence and their leaf nitrogen levels might be higher, and the perennials should produce these pigments and have lower leaf nitrogen levels at senescence. Similar scenarios can be envisaged for other functions, as herbivore defence or drought tolerance.

Finally, it is interesting to focus on the evolution of the genes responsible for the synthesis of anthocyanins in specific tissues. Patterns and rates of evolution may reveal much about the importance of these different gene products (Rausher *et al.*, 1999). In this long history of research on anthocyanin function in vegetative organs, we are now poised to solve some of the physiological and biochemical challenges and will then need to put this research in an evolutionary context.

REFERENCES

van Acker, S. A. B. E, van den Berg, D. -J., Tromp, M. N. J. L., Griffioen, D. H., van Bennekom, W. P., van der Vugh, W. J. F. and Bast, A. (1996). Structural aspects of antioxidant activity in flavonoids. *Free Radical Biology and Medicine* **20**, 331–342.

Archetti, M. (2000). The origin of autumn colours by coevolution. *Journal of Theoretical Biology* **205**, 625–630.

Bahler, B. D., Steffen, K. L. and Orzolek, M. D. (1991). Morphological and biochemical comparison of a purple-leafed and a green-leafed pepper cultivar. *HortScience* **26**, 736.

Barnes, J. (ed.) (1984). "The Complete Works of Aristotle", 2 volumes. Princeton University Press, Princeton, NJ, USA.

Beggs, C. J. and Wellman, E. (1994). Photocontrol of flavonoid biosynthesis. *In* "Photomorphogenesis in Plants" (R. E. Kendrick and G. H. M. Kronenberg, eds) pp. 733–751. Martinus Nijhoff, Dordrecht, The Netherlands.

Bors, W., Michel, C. and Saran, M. (1994). Flavonoid antioxidants: rate constants for reactions with oxygen radicals. *Methods in Enzymology* **234**, 420–429.

Brown, S. and Hamilton, W. (2001). Autumn leaf colours and herbivore defense. *Proceedings of the Royal Society of London B. Biological Sciences* **268**, 1489–1493.

Burchard, P., Bilger, W. and Weissenböck, G. (2000). Contribution of hydroxycinnamates and flavonoids to epidermal shielding of UV-A and UV-B radiation in developing rye primary leaves as assessed by ultraviolet-induced chlorophyll fluorescence measurements. *Plant Cell and Environment* **23**, 1373–1380.

Burger, J. and Edwards, G. E. (1996). Photosynthetic efficiency, and photodamage by UV and visible radiation in red versus green leaf *Coleus* varieties. *Plant Cell Physiology* **37**, 395–399.

Chalker-Scott, L. (1999). Environmental significance of anthocyanins in plant stress responses. *Photochemistry and Photobiology* **70**, 1–9.

Cittadino, E. (1990). "Nature as the Laboratory. Darwinian Plant Ecology in the German Empire, 1880–1900". Cambridge University Press, Cambridge.

Coley, P. D. and Aide, T. M. (1989). Red colouration of tropical young leaves: a possible antifungal defense? *Journal of Tropical Ecology* **5**, 293–300.

Coley, P. D. and Barone, J. A. (1996). Herbivory and plant defenses in tropical forests. *Annual Reviews of Ecology and Systematics* **27**, 305–335.

Costa-Arbulú, C., Gianoli E., Gonzáles W. and Niemeyer H. M. (2001). Feeding by the aphid *Sipha flava* produces a reddish spot on leaves of *Sorghum halepense*: an induced defence? *Journal of Chemical Ecology* **27**, 273–283.

Do, C. B. and Cormier, F. (1991). Accumulation of paeonidin-3-glucoside enhanced by osmotic stress in grape (*Vitis vinifera* L.) cell suspension. *Plant Cell Tissue and Organ Culture* **24**, 49–54.

Ewart, A. J. (1897). The effects of tropical insolation. *Annals of Botany* **11**, 439–480.

Feild, T. S., Lee, D. W. and Holbrook, N. M. (2001). Why leaves turn red in autumn. The role of anthocyanins in senescing leaves of red-osier dogwood. *Plant Physiology* **127**, 566–574.

Fineblum, W. L. and Rausher, M. D. (1997). Do floral pigmentation genes also influence resistance to enemies? The *W* locus in *Ipomoea purpurea*. *Ecology* **78**, 1646–1654.

Ganders, F. R., Griffiths, A. J. F. and Carey, K. (1980). Natural selection for spotted leaves: parallel morph ratio variation in 3 species of annual plants. *Canadian Journal of Botany* **58**, 689–693.

Geismann, T. A. and Swain, T. (1957). Biosynthesis of flavonoid compounds in higher plants. *Chemistry and Industry* **1957**, 984.

Givnish, T. J. (1987). Comparative studies of leaf form: assessing the relative roles of selective pressures and phylogenetic constraints. *New Phytologist* **106** (Suppl.), 131–160.

Givnish, T. J. (1990). Leaf mottling: relation to growth form and leaf phenology and possible role as camouflage. *Functional Ecology* **4**, 463–474.

Gould, K. S. and Quinn, B. D. (1999). Do anthocyanins protect leaves of New Zealand native species from UV-B? *New Zealand Journal of Botany* **37**, 175–178.

Gould, K. S., Kuhn, D. N., Lee, D. W. and Oberbauer, S. F. (1995). Why leaves are sometimes red. *Nature* **378**, 241–242.

Grew, N. (1682). "The Anatomy of Plants". W. Rawlins, London.

Grisebach, H. (1957). Zur biogenese des cyanidins 1. *Zeitschrift fuer Naturforschung* **12B**, 227–231.

Grisebach, H. and Patschke, L. (1961). Zur biogenese der Flavonoide. IV. 2′, 4, 4′, 6′-tetrahydroxy-chalkone-2′-glucosid-[b-14C] als Vorstufe fur Quercetin und Cyanidin. *Zeitschrift fuer Naturforschung* **16B**, 645–647.

Haberlandt, G. (1914). "Physiological Plant Anatomy", 4th edn. Macmillan and Company, London.

Hammaerschmidt, R. and Nicholson, R. L. (1977). Resistance of maize to anthracnose: effect of light intensity on lesion development. *Phytopathology* **67**, 247–250.

Harborne, J. B. (1976). Functions of flavonoids in plants. *In* "Chemistry and Biochemistry of Plant Pigments" (T. W. Goodwin, ed.), pp. 736–778. Academic Press, London.

Harborne, J. B. (1988). "The Flavonoids". Chapman and Hall, London.

Harborne, J. B. (1997). "Introduction to Ecological Biochemistry", 4th edn. Academic Press, New York.

Harborne, J. B. and Grayer, R. J. (1988). The anthocyanins. *In* "The Flavonoids. Advances in Research Since 1980" (J. B. Harborne, ed.), pp. 1–20. Chapman and Hall, London.

Hedin, P. A., Jenkins, J. N., Collum, D. H., White, W. H. and Parrott, W. L. (1983). Multiple factors in cotton contributing to resistance to the tobacco budworm, *Heliothis virescens* F. *In* "Plant Resistance to Insects" (P. A. Heden, ed.), pp. 347–367. American Chemical Society, Washington.

Hoagland, R. E. and Boyette, C. D. (1994). Pathogenic interactions of *Alternaria crassa* and phenolic metabolism in jimsonweed (*Datura stramonium* L.) varieties. *Weed Science* **42**, 44–49.

Hoch, W. A., Zeldin, E. L. and McCown, B. H. (2001). Physiological significance of anthocyanins during autumnal leaf senescence. *Tree Physiology* **21**, 1–8.

Juniper, B. (1993). Flamboyant flushes: a reinterpretation of non-green flush colours in leaves. *International Dendrological Society Yearbook*, 49–57.

Kakegawa, K., Hattori, E., Koike, K. and Takeda, K. (1991). Induction of anthocyanin synthesis and related enzyme activities in cell cultures of *Centaurea cyanus* by UV-light irradiation. *Phytochemistry* **30**, 2271–2273.

Keeble, F. W. (1895). The hanging foliage of certain tropical trees. *Annals of Botany* **9**, 59–93.

Kerner von Marilaum, A. (1897). "The Natural History of Plants" (translated by F. W. Oliver). Blackie, London.

Klaper, R., Frankel, S. and Berenbaum, M. R. (1996). Anthocyanin content and UVB sensitivity in *Brassica rapa. Phytochemistry* **63**, 811–813.

Krol, M., Gray, G. R., Hurry, M. V., Öquist, G., Malek, L. and Huner, N. P. A. (1995). Low-temperature stress and photoperiod affect an increased tolerance to photoinhibition in *Pinus banksiana* seedlings. *Canadian Journal of Botany* **73**, 1119–1127.

Kursar, T. A. and Coley, P. D. (1992). Delayed development of the photosynthetic apparatus in tropical rain forest species. *Functional Ecology* **6**, 411–422.

Lambers, H., Chapin III, F. S. and Pons, T. L. (1998). "Plant Physiological Ecology". Springer Verlag, New York.

Lee, D. W. and Collins, T. H. (2001). Phylogenetic and ontogenetic influences on the distribution of anthocyanins and betacyanins in leaves of tropical plants. *International Journal of Plant Science* **162**, 1141–1153.

Lee, D. W., Lowry, J. B. and Stone, B. C. (1979). Abaxial anthocyanin layer in leaves of tropical rain forest plants: enhancer of light capture in deep shade. *Biotropica* **11**, 70–77.

Lee, F. W., Brammeier, D. and Smith, A. P. (1987). The selective advantages of anthocyanins in developing leaves of mango and cacao. *Biotropica* **19**, 40–49.

Lee, D. W., O'Keefe, J., Holbrook, N. M. and Feild, T. S. (2002). Pigment dynamics and autumn leaf senescence in a New England deciduous forest. Submitted for publication.

Leyva, A., Jarillo, J. A., Salinas, J. and Martinez-Zapater, J. M. (1995). Low temperature induces the accumulation of phenylalanine ammonia-lyase and chalcone synthase mRNAs of *Arabidopsis thaliana* in a light-dependent manner. *Plant Physiology* **108**, 39–46.

Li, J., Ou-Lee, T. -M., Raba, R., Amundson, R. G. and Last, R. L. (1993). *Arabidopsis* flavonoid mutants are hypersensitive to UV-B irradiation. *The Plant Cell* **5**, 171–179.

Lois, R. and Buchanan, B. B. (1994). Severe sensitivity to ultraviolet radiation in an *Arabidopsis* mutant deficient in flavonoid accumulation. II. Mechanisms of UV-resistance in *Arabidopsis. Planta* **194**, 504–509.

Mancinelli, A. L. (1985). Light dependent anthocyanin synthesis: a model system for the study of plant morphogenesis. *Botanical Review* **51**, 107–157.
Marquart, L. C. (1835). "Die Farben der Blüthen, Eine chemisch-physiologische Abhandlung". Habicht, Bonn.
McClure, J. W. (1975). Physiology and function of flavonoids. *In* "The Flavonoids" (J. B. Harborne, T. J. Mabry and H. Mabry, eds) pp. 970–1055. Chapman and Hall, London.
McKown, R., Kuroki, G. and Warren, G. (1996). Cold responses of *Arabidopsis* mutants impaired in freezing tolerance. *Journal of Experimental Botany* **47**, 1919–1925.
Mita, S., Murano, N., Akaike, M. and Nakamura, K. (1997). Mutants of *Arabidopsis thaliana* with pleiotropic effects on the expression of the gene for beta-amylase and on the accumulation of anthocyanin that are inducible by sugars. *Plant Journal* **11**, 841–851.
Mohl, H. von (1837). "Untersuchungen über die winterliche Färbung der Blätter. Vermischte Schriften botanischen inhalts". G. Bähr, Tübingen.
Mol, J., Jenkins, G. I., Shafer, E. and Weiss, D. (1996). Signal perception, transduction, and gene expression involved in anthocyanin biosynthesis. *Critical Reviews in Plant Science* **15**, 525–557.
Murray, J. R., Smith, A. G. and Hackett, W. P. (1994). Differential dihydroflavonol reductase transcription and anthocyanin pigmentation in the juvenile and mature phases of ivy (*Hedera helix* L.) *Planta* **194**, 102–109.
Neill, S. O. (2002). The functional role of anthocyanins in leaves. Ph.D. thesis, University of Auckland, New Zealand.
Neill, S. O. and Gould, K. S. (1999). Optical properties of leaves in relation to anthocyanin concentration and distribution. *Canadian Journal of Botany* **77**, 1777–1782.
Neill, S. O., Gould, K. S., Kilmartin, P. A., Mitchell, K. A. and Markham, K. R. (2002). Antioxidant activities of red versus green leaves in *Elatostema rugosum. Plant Cell and Environment* **25**, 537–549.
Nicholson, R. L., Kollipara, S. S., Vincent, J. R., Lyons, P. C. and Cadena-Gomez, G. (1987). Phytoalexin synthesis by the sorghum mesocotyl in response to infection by pathogenic and non-pathogenic fungi. *Proceedings of the National Academy of Sciences, USA* **84**, 5520–5524.
Onslow, M. S., ed. (1926). "The Anthocyanin Pigments of Plants". University of Cambridge Press, Cambridge.
Pick, H. (1883). Ueber die Bedeutung des rothen Farbstoffes bei den Phanerogamen und die Beziehungen desselben zur Stärkewanderung. *Botanisches Centralblatt fuer Deutschland* **116**, 281–382.
Pringsheim, N. (1879-1882). Ueber Lichtwirkung und Chlorophyllfunction in der Pflanze. *Jahrbuch fuer Wissenschaftliche Botanik,* Volumes 12/13, G. Boratrager, Berlin.
Rausher, M. D., Miller, R. E. and Tiffin, P. (1999). Patterns of evolutionary rate variation among genes of the anthocyanin biosynthetic pathway. *Molecular Biology and Evolution* **16**, 266–274.
Rice-Evans, C. A. and Packer, L., eds (1998). "Flavonoids in Health and Disease". Marcel Dekker, New York.
Rosenthal, G. A. and Janzen, D. (1979). "Herbivores: Their Interactions with Secondary Plant Metabolites". Academic Press, New York.
Sage, L. C. (1992). "Pigment of the Imagination. A History of Phytochrome Research". Academic Press, San Diego.
Sherwin, H. W. and Farrant, J. M. (1998). Protection mechanisms against excess light in the resurrection plants *Craterostigma wilmsii* and *Xerophyta viscose. Plant Growth Regulation* **24**, 203–210.
Shirley, B. W. (1996). Flavonoid biosynthesis: 'new' functions for an 'old' pathway. *Trends in Plant Science* **1**, 377–382.

Simms, E. L. and Bucher, M. A. (1996). Pleiotropoic effects of flower color intensity on herbivore performance in *Ipomoea purpurea*. *Evolution* 50, 957–963.

Smillie, R. M. and Hetherington, S. E. (1999). Photoabatement by anthocyanin shields photosynthetic systems from light stress. *Photosynthetica* **36**, 451–463.

Smith, A. M. (1909). On the internal temperature of leaves in tropical insolation, with special reference to the effect of their colour on temperature: also observations on the periodicity of the appearance of young coloured leaves growing in the Peradeniya Gardens. *Annals of the Royal Botanical Garden, Peradeniya* **4**, 229–298.

Smith, A. P. (1986). Ecology of a leaf colour polymorphiyism in a tropical forest species: habitat segregation and herbivory. *Oecologia* **69**, 283–287.

Smith, H. (1994). Sensing the light environment: the functions of the phytochrome family. *In* "Photomorphogenesis in Plants" (R. E. Kendrick and G. H. M. Kronenberg, eds), pp. 377–416. Martinus Nijhoff, Dordrecht, The Netherlands.

Stafford, H. A. (1990). "Flavonoid Metabolism". CRC Press. Boca Raton, FL, USA.

Stahl, E. (1896). Ueber bunte Laublätter. *Annals of the Botanical Garden Buitenzorg* **13**, 137–216.

Stapleton, A. E. and Walbot, V. (1994). Flavonoids can protect maize DNA from the induction of ultraviolet radiation damage. *Plant Physiology* **105**, 881–889.

Stiles, E. W. (1982). Fruit flags: two hypotheses. *American Naturalist* **120**, 500–509.

Stone, B. C. (1979). Protective colouration of young leaves in certain Malaysian palms. *Biotropica* **11**, 126.

Sturgeon, K. B. and Mitten, J. B. (1980). Cone colour polymorphism associated with elevation in white fir, *Abies concolor*, in southern Colorado. *American Journal of Botany* **67**, 1040–1045.

Takahashi, A., Takeda, K. and Ohnishi, T. (1991). Light-induced anthocyanin reduces the extent of damage to DNA in UV-irradiated *Centaurea cyanus* cells in culture. *Plant Cell Physiology* **32**, 541–547.

Theophrastus (1918). "Enquiry into Plants", Volumes 1 and 2 (translated by A. Hort). Harvard University Press, Cambridge, USA.

Tholakalabavi, A., Zwiabek, J. J. and Thorpe, T. A. (1997). Osmotically-stressed poplar cell cultures: anthocyanin accumulation, deaminase activity, and solute composition. *Journal of Plant Physiology* **151**, 489–496.

Watkin, J. E., Underhill, E. W. , and Neish, A. C. (1957). Biosynthesis of quercetin in buckwheat I. *Canadian Journal of Biochemistry and Physiology* **35**, 219–228.

Westhoff, P. (1998). "Molecular Plant Development". Oxford University Press, Oxford.

Wheldale, M. (1916). "The Anthocyanin Pigments of Plants". Cambridge University Press, Cambridge.

Willstätter, R. and Everest, A. E. (1912). Ueber den Farbstoff der Kornblume. *Liebigs Annalen der Chemie* **101**, 189–232.

Willstätter, R. and Mallison, H. (1914). Ueber die Verwandtschaft der Anthocyane und Flavone. *Sitzung berichte der Deutschen Akademie der Wissenschaften zu Berlin* **34**, 769–777.

Woodall, G. S. and Stewart, G. R. (1998). Do anthocyanins play a role in UV protection of the red juvenile leaves of *Syzygium*? *Journal of Experimental Botany* **325**, 1447–1450.

Yamasaki, H. (1997). A function of colour. *Trends in Plant Science* **2**, 7–8.

Le Rouge et le Noir: Are Anthocyanins Plant Melanins?

GRAHAM S. TIMMINS[1], N. MICHELE HOLBROOK[2] and TAYLOR S. FEILD[3]

[1]College of Pharmacy, University of New Mexico, Albuquerque, NM 87131, USA
[2]Department of Organismic and Evolutionary Biology, Harvard University, Cambridge, MA 02138, USA
[3]Department of Integrative Biology, University of California, Berkeley, CA 94720-3140, USA

ABSTRACT

Anthocyanin and melanin are important determinants of coloration in plants and animals. Although chemically unrelated, anthocyanin and melanin share many physiological and structural traits, as well as exhibit some important differences. Of the two, melanin has been much more extensively studied and its role as both a photoprotectant and free radical scavenger is relatively well established. In contrast, the physiological properties of anthocyanin have received less attention. However, like melanin anthocyanin appears to play a dual role of both attenuating light intensities within superficial (light-penetrated) tissues and functioning as an antioxidant.

Advances in Botanical Research Vol. 37
incorporating Advances in Plant Pathology
ISBN 0-12-005937-1

I. INTRODUCTION

Coloration in both plants and animals plays an important role in guiding interactions with organisms capable of detecting such colors. In plants, anthocyanins are the principal pigment used in both attracting (pollinators, dispersal agents) and evading (herbivores) animals (Wheldale, 1916), while in animals, melanin pigments play a similar role in terms of surface coloration (skin, eyes, hair), with consequent effects on mate attraction, aggressive displays, and camouflage (Fox and Vevers 1960). The existence of these two pigment systems leads us to ask whether these similarities are only 'skin-deep'. Specifically, do anthocyanins and melanins share analogous functional roles beyond their involvement in external coloration? Melanins have been the focus of extensive study and their physiological effects, in terms of photoprotection and quenching of free radicals, are reasonably well understood (Jimbow *et al.*, 1986; Riley, 1997). In contrast, plant anthocyanins have received much less attention and their functional properties are only beginning to be explored (Feild *et al.*, 2001; Neill, 2002). Yet there is increasing evidence that anthocyanins may possess similar physiological effects to those demonstrated for melanin. In this essay we compare and contrast the structure and physiology of anthocyanins and melanins. Our goal, however, is not to equate these two pigment systems, but rather to explore how divergent lineages have utilized pigmented molecules to resolve physiological problems associated with the penetration of light into their superficial layers.

II. STRUCTURE AND OCCURRENCE

A. ANTHOCYANINS

Anthocyanin pigments occur in all major groups of land plants, including seed plants, ferns and bryophytes (Lee *et al.*, 1987; Post, 1990; Post and Meret, 1992; Kunz *et al.*, 1994; Krol *et al.*, 1995; Shirley, 1996). They are expressed *par excellence* in the flowering plants, where they are often responsible for the pleasing colors of flowers and fruits and function in the attraction of both pollinators and dispersal agents (McClure, 1975;

Harborne, 1988b). However, anthocyanins are not limited to reproductive structures and can occur in a variety of vegetative tissues, including leaves, leaf veins, young stems, and even roots. Indeed, the advent of fiery-red and wine-purple leaves during autumn senescence is one of the best known and economically important temporal displays of anthocyanins (Feild *et al.*, 2001; Hoch *et al.*, 2001; Lee *et al.*, 2002). Anthocyanins can also occur in actively growing tissues, for example the flushing leaves of mango and cacao (Lee *et al.*, 1987; Woodall and Stewart, 1988; Juniper, 1994). The presence of anthocyanins in vegetative tissues raises the question of whether they might have a physiological or an ecological function (Lee and Collins, 2001; Neill, 2002).

Anthocyanins are water-soluble pigments synthesized in the cytosol and subsequently transported into the vacuole (Hrazdina *et al.*, 1978; Harborne and Grayer, 1988; Marrs *et al.*, 1995). Within leaves, they are most commonly found in the vacuoles of palisade and spongy parenchyma, however, in some species they can be found within the lower or upper epidermal layers (Neill and Gould, 1999; Gould *et al.*, 2000; Lee *et al.*, 2002). Anthocyanin concentrations are typically closely correlated with light intensity (Grace *et al.*, 1998; Chalker-Scott, 1999; Feild *et al.*, 2001; Hoch *et al.*, 2001; Lee *et al.*, 2002), suggesting that their synthesis is stimulated by light (Beggs and Wellmann, 1985; Mol *et al.*, 1996). Anthocyanin production is often enhanced by stress (Harborne, 1988a; Chalker-Scott, 1999), such as temperature extremes (Christie *et al.*, 1994; Pictrini and Massacci, 1998), low nutrient availability (Bongue-Bartelsman and Phillips, 1995; Deldaldechamp *et al.*, 1995; Trull *et al.*, 1997), UV-radiation (Mendez *et al.*, 1999), lack of water (Balakumar *et al.*, 1993; Sherwin and Farrant, 1998), mechanical injury (Ferreres *et al.*, 1997), and herbivores and pathogens (Dixon *et al.*, 1994).

The anthocyanin molecule consists of an anthocyanidin (the aglycone chromophore) bonded to one or more glycosides (Harborne and Grayer, 1988). The chromophore has a C6-C3-C6 configuration consisting of two aromatic rings, connected by a heterocyclic ring (Fig. 1). Although there are 18 known naturally occurring anthocyanidins, the most common anthocyanin in leaves is cyanidin-3-glucoside (Harborne, 1967). The high degree of conjugation of the three planar rings, coupled with the relatively low degree of saturation, contributes to the ability of anthocyanins to absorb appreciable amounts of visible radiation (Harborne, 1967; Shirley, 1996). Anthocyanins are unique among the flavonoids in their ability to absorb light within the visible spectrum (Harborne, 1967). However, their capacity to absorb light within the ultraviolet range is correspondingly reduced relative to other, less modified flavonoids (Shirley, 1996). Anthocyanins have peak absorbance in the green-yellow region of the spectrum (500–550 nm), with a smaller peak in the ultraviolet

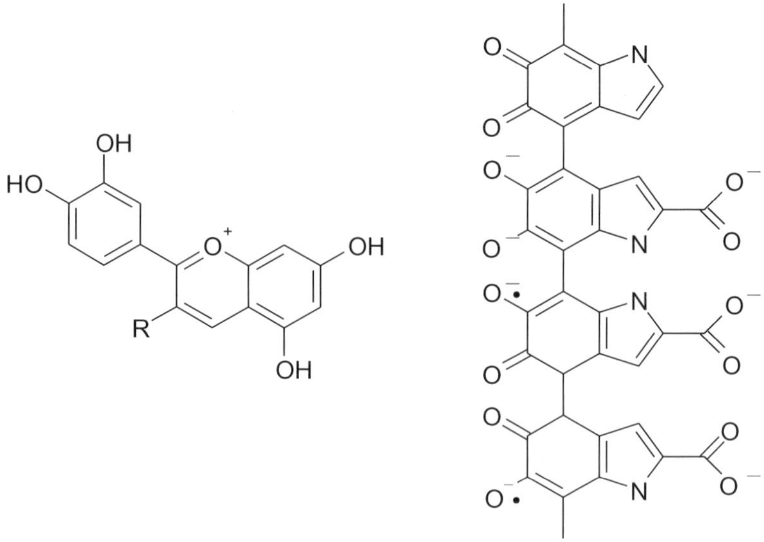

Fig. 1. Chemical structure of anthocyanin and (eu)melanin.

(Harborne, 1967). They appear ‘red’ largely because of their removal of green and yellow wavelengths, rather than from a failure to absorb red wavelengths (Neill, 2002). Anthocyanins also can take on a variety of colors, ranging from pink to blue, due largely to the degree of substitution, local pH, co-pigmentation with other flavonoids, and chelation to metal ions (Brouillard and Dangles, 1993; Markham *et al.*, 2000).

B. MELANINS

Melanin is a black pigment that plays an important role in determining skin and hair coloration in many animals (Fox and Vevers, 1960; Jimbow *et al.*, 1986; Hill, 1992). A dark form of the pigment, eumelanin, occurs in a range of animal tissues, including feathers, fur, hair, beaks, scales, and skin. The ink of octopus and squid is a suspension of melanin, and was the original source for sepia. However, melanins are not always black; the sulfur-containing forms, known as pheomelanins, are responsible for ‘red’ hair in mammals, as well as reddish-brown and some yellow feathers in birds (Fox and Vevers, 1960; Jimbow *et al.*, 1986).

In vertebrates, melanin is synthesized in membrane-bound organelles called melanosomes (Nordlund and Boissy, 2001). Melanosomes become activated in melanogenic cells that make up the retinal pigment epithelium and in neural crest-derived dendritic cells known as melanocytes. In contrast to anthocyanins, melanin producing cells are not limited to external, light-penetrated tissues (Nordlund and Boissy, 2001). Most melanocytes, however, occur in the skin, although significant numbers occur in both the eyes and the ears (Nordlund and Boissy, 2001). Epidermal melanocytes migrate to their final location in the basal layer of the epidermis with their branches extending between the Malpighian cells (Jimbow *et al.*, 1986). Melanin granules are produced by melanocytes and delivered to hair and epidermal cells by the ends of the dendritic processes which become incorporated into the epithelial cells (Wolff *et al.*, 1974; Riley, 1997). Within these cells, the melanosomes are concentrated on the apical surface where they appear to form a 'cap' over the nucleus (Farrington, 1964; Riley, 1997; Kobayashi *et al.*, 1998). Differences in skin color are due to larger and more numerous melanosomes, with the number of melanocytes being relatively constant (Nordlund and Boissy, 2001).

Melanins are a complex and somewhat heterogeneous group of polymers, whose exact chemical structures are not known (Riley, 1997; Meyskens *et al.*, 2001). Most melanins appear to be mixed polymers based on indoles (primarily 4′-7 linked substituted indoles; Fig. 1), but which contain a variable amount of other pre-indolic products (Riley, 1997; Prota, 2000). Melanins are synthesized from the oxidation of L-tyrosine, resulting in a conjugated polymer with the capacity for electronic reactions, photon absorption, and cation binding (Riley, 1997). Pheomelanins result from the incorporation of cystein (Jimbow *et al.*, 1986). The high degree of conjugation is responsible for the ability of melanin to absorb broadly across the visible spectrum (as well as in the ultraviolet), resulting in the dark color of this pigment (Riley, 1997). Melanins can take place in both one- and two-electron redox reactions, resulting in the formation of a semiquinone radical intermediate. Anthocyanins can also be oxidized to such semiquinone radical intermediates and so we postulate that similarities in molecular structure derive from similarities in molecular action. The presence of carboxyl groups is responsible for melanins being strong cation chelators, allowing them to bind potentially toxic cations such as transition metals (Riley, 1997; Meyskens *et al.*, 2001).

Like anthocyanins, melanin synthesis is stimulated by light, although in this case the critical part of the spectrum is in the ultraviolet (Quevedo *et al.*, 1975; Pathak and Fanselow, 1983; Kollias *et al.*, 1991; Nordlund and Boissy, 2001). Exposure to ultraviolet radiation leads to darker skin coloration due to three mechanisms (Farrington, 1964). First, existing melanin molecules become photo-oxidized, thus increasing their carbonyl content, which both increases the ability of melanin to absorb light and results in a rapid increase in skin coloration known as the immediate pigment darkening reaction. The second mechanism of 'tanning' is the migration of pre-formed melanin to upper stratum Malpighii cells and the stratum corneum. Finally, ultraviolet radiation stimulates the enzyme tyrosinase, which catalyzes the key initial step in melanin synthesis (Jimbow *et al.*, 1986). The recessive mutation that results in albinism is due to a lack of this enzyme. Repeated exposure to ultraviolet light leads to an increase in the number of functioning (i.e., melanosome producing) melanocytes (Jimbow *et al.*, 1986). Interestingly, melanin synthesis in hair or feathers is not increased by light (Fox and Vevers, 1960). Melanin synthesis is also affected by a number of 'melanogenic' hormones (Nordlund and Boissy, 2001) and, in humans, can be stimulated by exposure to cold (Fox and Vevers, 1960).

III. PHOTOPROTECTION

Light is a source of information and energy to both plants and animals and thus it is not surprising that each type of organism contains surfaces designed to make use of incident radiation. We typically think of light as primarily an energy source for plants given its central role in powering photosynthesis. However, plants also use light to measure daylength and to assess their local environment, such as their proximity to neighboring plants (Gilbert *et al.*, 2001). In contrast, we generally think of animals as primarily using light to gain information about their environment with the most developed form of this being the ability to form visual 'images' of their surroundings. However, absorption of light energy plays an important role in the thermal balance of many animals and some even use absorbed radiation to drive chemical reactions (e.g., the synthesis of vitamin D in mammals). Because of this shared dependence on light, both animals and plants are susceptible to damage that can result from excessive amounts of solar radiation.

In both plants and animals, absorption of ultraviolet light can lead to cellular damage resulting from dimerization of nucleic acids (Mitchell, 1995). Excessive intensities of visible radiation can also induce cellular damage by photo-oxidation, resulting from light-induced stimulation of reactive oxygen species (ROS). Two basic 'strategies' for minimizing the physiological impacts of light-induced stress are to prevent the occurrence of damaging light intensities (suppressing or avoidance mechanisms) or to invest in mechanisms that respond to potential downstream sequences of events that are triggered by light-driven damage (scavenging or tolerance mechanisms; Asada, 1999). Because of their ability to strongly capture visible light, both melanin and anthocyanins have the potential to act as photoprotective pigments that block light before it can damage cells. Below we review the evidence that these two pigment systems act as natural sunscreens in both plants and animals.

A. ANTHOCYANINS

Excess light intensities, defined as when the rate of photon absorption exceeds the rate of photon utilization by photosynthesis, can set in motion processes that significantly damage the photosystem II (PSII) reaction center as well as membranes and proteins by singlet oxygen ($^1O_2$), superoxide (O_2^-), and other free radicals that are sensitized by light. Specifically, over-stimulation of the light reactions can increase the rate of chlorophyll triplet state formations in PSII, which can, in turn, transfer their energy to molecular oxygen to form highly reactive singlet oxygen (Asada, 1999). Singlet oxygen has a short life time in aqueous media, however, it can damage cells either by reacting directly with the PSII reaction center or by oxidizing chloroplast membrane lipids which compromises their ability to maintain ionic gradients (Asada, 1994). In addition, excessive amounts of reduced ferredoxin, formed on the reducing side of photosystem I, can donate electrons to O_2, resulting in the production of superoxide as $NADP^+$ becomes limiting when the rate of light absorption outpaces the rate of Rubisco turnover (Asada, 1999).

Thus, there are numerous mechanisms and sites of vulnerability of the photosynthetic apparatus that can lead to its 'self-destruction' under excess light duress. Not surprisingly, plants have evolved an array of biochemical defenses against excess light energy, ranging from processes that harmlessly dissipate the excess photons as heat and others that enable the photosynthetic electron transport chain to continue turnover even in the face of low CO_2 availability or low Rubisco capacity (Demmig-Adams and Adams, 1992; Asada, 1999). However under conditions of stress or developmental alteration of the photosynthetic membranes, the biochemical safety valves that normally protect the chloroplast can become disrupted, and anthocyanins represent another photoprotective process that steps in under these circumstances. In contrast to visible light, photosynthetic tissues can not utilize ultraviolet light to drive chloroplast electron transport enabling leaves to shield themselves from this potentially damaging part of the spectrum through the deployment of ultraviolet light-absorbing flavonoids in their upper, epidermal cell layer (Caldwell *et al.*, 1983; Day, 1993; Koes *et al.*, 1994).

The idea that anthocyanins could function as a photoprotectant dates back to the 19th century (Wheldale, 1916). However, perhaps because anthocyanins are sequestered in cell vacuoles, and therefore physically disjoined from chloroplasts, their effects in altering the light climate within leaves has, until recently (Gould *et al.*, 1995; Krol *et al.*, 1995; Pietrini and Massacci, 1998; Sherwin and Farrant, 1998; Neill and Gould, 1999; Smillie and Hetherington, 1999; Feild *et al.*, 2001; Hoch *et al.*, 2001; Neill, 2002), received little attention (Lee *et al.*, submitted). Yet, anthocyanins frequently occur near the sun-facing surfaces of leaves

(Feild *et al.*, 2001; Lee *et al.*, submitted) and are effective at attenuating the wavelengths of light most able to penetrate deep within photosynthetic tissues (green and yellow wavelengths; Nishio, 2000), making them potentially effective light-screening pigments.

Anthocyanins are most common in tissues that are either newly expanding, senescing, or experiencing environmental stress (Chalker-Scott, 1999; Lee and Collins, 2001). In the first two cases, the 'need' for a photoprotective mechanism above-and-beyond that of other pigments directly involved in chloroplast-level photoprotection (e.g., xanthophyll carotenoids, see Demmig-Adams and Adams, 1992 for a review) may reflect the relative impairment of photoprotective processes during chloroplast development when these are not fully on-line, or during chloroplast senescence when the functioning of the xanthophyll cycle may become disrupted by membrane degradation (Feild *et al.*, 2001). Stressful conditions, such as cold temperatures or low nutrient availability, decrease the capacity for carbon assimilation and thus may alter the light levels that can be safely processed by the chloroplast (Demmig-Adams and Adams, 1992; Asada, 1999). In all of these situations, photosynthetic rates become limited by substrates other than light absorption and thus the production of a photoprotectant that reduces light intensities incident on the chloroplasts forms an appropriate mechanism for avoiding damage.

A number of recent studies provide evidence that anthocyanins function as sunscreens in mature leaves (Sharma and Banerji, 1981; Gould *et al.*, 1995; Krol *et al.*, 1995; Sherwin and Farrant, 1998; Close *et al.*, 2000; Grace and Logan, 2000; Neill, 2002), senescing leaves (Feild *et al.*, 2001), developing fruits (Smillie and Hetherington, 1999; Merzlyak and Chivkunova, 2000), and expanding leaves (Baker and Hardwick, 1973; Lee *et al.*, 1987). The common feature of these studies is the demonstration of reduced photoinhibition in tissues that contain anthocyanins compared to those that lack anthocyanins.

As an example of anthocyanins acting as photoprotectants, Feild *et al.*, (2001) examined the role of anthocyanins as screening pigments during autumn leaf senescence in red-osier dogwood (*Cornus stolonifera* L.). Before chlorophyll is appreciably broken down, dogwood leaves exposed to full sunlight synthesize anthocyanins in the palisade layer and shift from summer-green to autumn-purple. In contrast, *Cornus* leaves from shadier environments lack detectable amounts of anthocyanins and turn yellow as carotenoids are unmasked by declining chlorophyll concentrations during senescence. The system provided a good comparison, as there were no significant differences in the concentrations of chlorophyll or total carotenoids and anatomical characteristics between sun and shade *Cornus* leaves (Feild *et al.*, 2001). In addition, light response curves of photosystem II quantum yield made on the leaf undersurface, thus avoiding the light screening effect of the anthocyanic layer, demonstrated that the chloroplasts within the spongy mesophylls of both leaf types had similar photosynthetic properties. The photoprotective role of anthocyanins in these senescing leaves was seen in the ability of PSII quantum yield of anthocyanic leaves to recover more quickly and to a greater extent following a high light dose (at an intensity equivalent to full sunlight during fall) in comparisons to yellow-senescing leaves of similar age (Fig. 2A). The effect of this excess light treatment was greatest when leaves were challenged with blue light, which is intensely captured by anthocyanins, and was greatly reduced when the leaves were treated with the same fluence of red light, which is poorly absorbed by anthocyanins (Fig. 2B, C). Although the study of the optical effects of anthocyanins in plants under natural circumstances have produced significant advances, future efforts aimed at understanding the photoprotective functions of anthocyanins should make use of more genetically controlled systems. For example, studies on the photoinhibitory responses of *Arabidopsis* anthocyanin-less mutants as well as woody plant cultivars that constitutively express or completely lack anthocyanins would be useful in examining the roles that anthocyanins play in protection of the photosynthetic apparatus during its construction as leaves expand, breakdown during leaf senescence, and protection under stress.

B. MELANINS

Numerous studies have demonstrated that protection from damaging effects of solar radiation is directly correlated with melanin content (Azizi *et al.*, 1988; Kollias *et al.*, 1991; Morison, 1995). However, the role of melanins as photoprotectants is perhaps best illustrated by their commercial inclusion in both sunscreen lotions and sun-glasses (Cesarini and Msika, 1995). Melanins absorb broadly across the ultraviolet and visible spectrum and can safely dissipate this energy as heat (Hill, 1992; Riley, 1997). For animals, the major source of damage from the solar spectrum is in the ultraviolet, which can delay or inhibit cell division and directly damage DNA (Farrington, 1964). However, it is also here that the absorption of radiation of less than 310 nm by Malpighian cell layers is directly utilized in the synthesis of vitamin D3 (Neer, 1975). Deeply pigmented human skin absorbs about 50% of incident sunlight (Nordlund and Boissy, 2001) and thus can influence rates of vitamin D3 synthesis (Matsuoka *et al.*, 1991; Rostand, 1997). However, the conflict between photoprotection and utilization in animals is not as great as in plants and it is unlikely that even very high levels of melanin in humans are sufficient to prevent the generation of normal levels of vitamin D synthesis (Matsuoka *et al.*, 1991; Nordlund and Boissy, 2001).

A

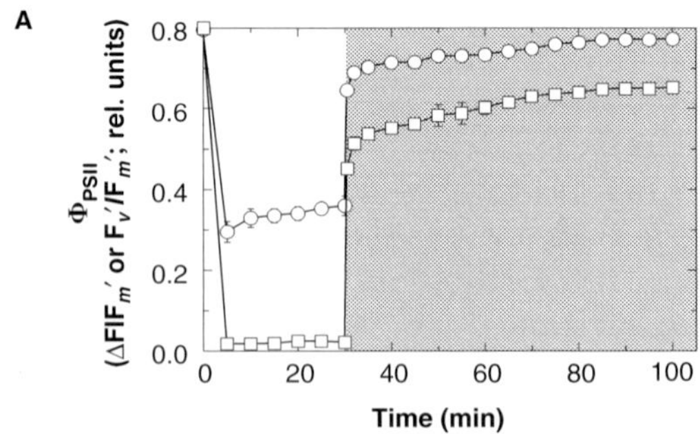

B

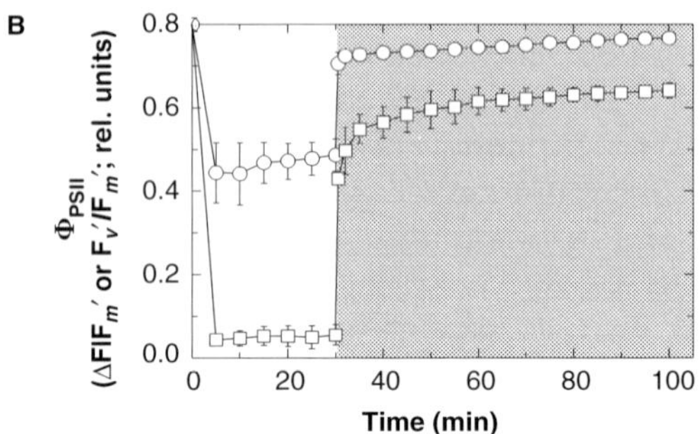

C

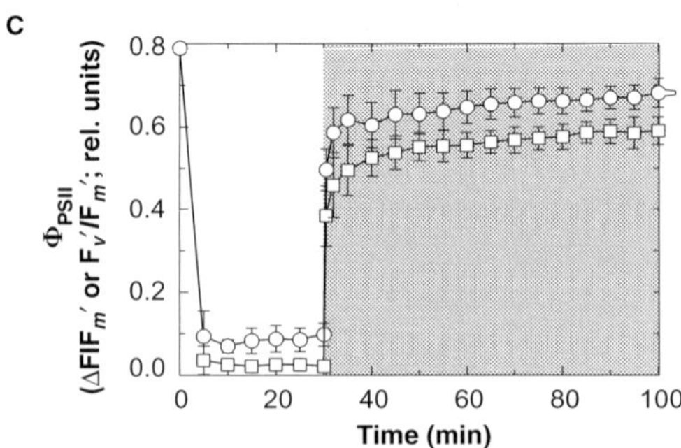

Fig. 2. Changes in effective PSII photon efficiency (Φ_{PSII} under illumination and F_v'/F_m' following darkening) to excess light intensity (1500 ± 50 μmol photons m^{-2} s^{-1} light) treatments of varying wavelength distribution [A, illuminated with white (400–800 nm) light; B, illuminated with blue (400–550 nm) light; and C, illuminated with red (640–710 nm) light] for red- (circles) and yellow-senescing (squares) *Cornus stolonifera* leaves. The light was turned off after 30 minutes (as indicated by the shaded box) and Φ_{PSII} recovery measured using a pulse-amplitude modulated fluorometer (PAM-2000, Heniz Walz, Germany). Measurements were made on detached leaves in a humidified chamber at constant gas concentration (380 μl l^{-1} carbon dioxide, 21% oxygen balanced with nitrogen gas) and temperature (20 ± 2 °C). Each curve for panels A–C is an average of five leaves per treatment and error bars denote the standard deviation. Reprinted from Feild *et al.* (2001).

IV. ANTIOXIDANT CAPACITY

Tissues exposed to sunlight are prone to the formation of reactive oxygen species (ROS) – a class of compounds that includes the oxygen radicals (e.g., superoxide, hydroxyl, and peroxyl radicals), as well as non-radicals such as singlet oxygen ($^1O_2$), hydrogen peroxide (H_2O_2), and ozone (O_3). All of these strongly oxidizing agents are capable of causing significant cellular damage, through their interactions with nucleic acids, proteins, and lipids (Foyer *et al.*, 1994; Alscher *et al.*, 1997; Polle, 1997). In photosynthetic tissues, the major mechanism by which energy (either directly or in the form of high-energy electrons) is passed to oxygen is via the light absorbing properties of chlorophyll (Foyer *et al.*, 1994). Limitations on the ability to safely utilize this absorbed radiation can lead to increased energy transfer to oxygen. In animals, it is the ultraviolet region of the spectrum that is responsible for the generation of ROS (Meyskens *et al.*, 2001). Furthermore, in animals, irradiation of melanin can lead to the production of ROS (Hill, 1992), although the presence of other chromophores in skin makes it difficult to determine the extent to which melanin is responsible for the generation of free radicals.

Although ROS are the nearly inevitable consequence of many cellular activities, their damaging effects can be minimized through the action of 'antioxidants'. Antioxidants defuse ROS through their ability to either donate or accept an electron and in this way 'scavenge' free radicals, or by direct energy transfer which can 'quench' the dangerous compound. It is becoming increasingly evident that polyphenols play important roles as antioxidants in terms of their ability to scavenge free radicals. Anthocyanins are reported to scavenge a wide variety of ROS including H_2O_2 and O_2^- (Bors *et al.*, 1990; Chauhan *et al.*, 1992; Bors *et al.*, 1994; Yamasaki *et al.*, 1996; Yamasaki, 1997; Tsuda *et al.*, 2000b). Recent work on the antioxidant function of anthocyanins has focused upon their dietary uses as antioxidants (Tsuda *et al.*, 1998; Kimura *et al.*, 1999; Tsuda *et al.*, 1999a; Tsuda *et al.*, 1999b; Igarashi *et al.*, 2000; Tsuda *et al.*, 2000a; Hagiwara *et al.*, 2001; Ramirez-Tortosa *et al.*, 2001), and also upon *in vitro* assays of the activity of various anthocyanins (Tsuda *et al.*, 1996; Abuja *et al.*, 1998; Kaneyuki *et al.*, 1999). Similarly, the polyphenolic nature of melanin has led to speculation that in addition to its light absorbing function, it can also scavenge UV-generated radicals (Bustamante *et al.*, 1993; Blarzino *et al.*, 1999; Rozanowska *et al.*, 1999; Meyskens *et al.*, 2001). The similarity of these two molecular mechanisms of action is shown in Fig. 3, demonstrating how analogous resonance stabilized radical species can result from the reactions of anthocyanins and melanins with radicals.

Collins *et al.* (1995) demonstrated that stable semiquinone-radicals (Fig. 3) can be directly observed by electron paramagnetic resonance

(EPR) in UV-irradiated skin, as their low reactivity makes them relatively persistent. These UV-induced melanin radicals indicate that melanin is indeed acting as an antioxidant in skin tissue, scavenging UV-induced free radical species to achieve a biologically-protective effect. We postulated that it might be possible to detect the entirely analogous protective effect of anthocyanins by irradiating red and green *Cornus stolonifera* leaves with visible light *in situ* within an EPR spectrometer and observing the semiquinone radical formed in the red leaves. Figure 4 shows the results of our preliminary experimentation, which suggests that a light-dependent semiquinone-like radical, similar to that seen for chemical oxidation of anthocyanins (Sakihama *et al.*, 2000) could be directly detected upon light exposure. This species was relatively persistent (as was the case for the melanin radical in skin), and was much greater in red than green senescing leaves, supporting our interpretation that it is anthocyanin derived. These data support our concept that the molecular similarity between anthocyanins and melanins may well derive in part from their similar combinations of light-screening and antioxidative functions, and we are currently studying anthocyanin-deficient mutants to further support our initial findings.

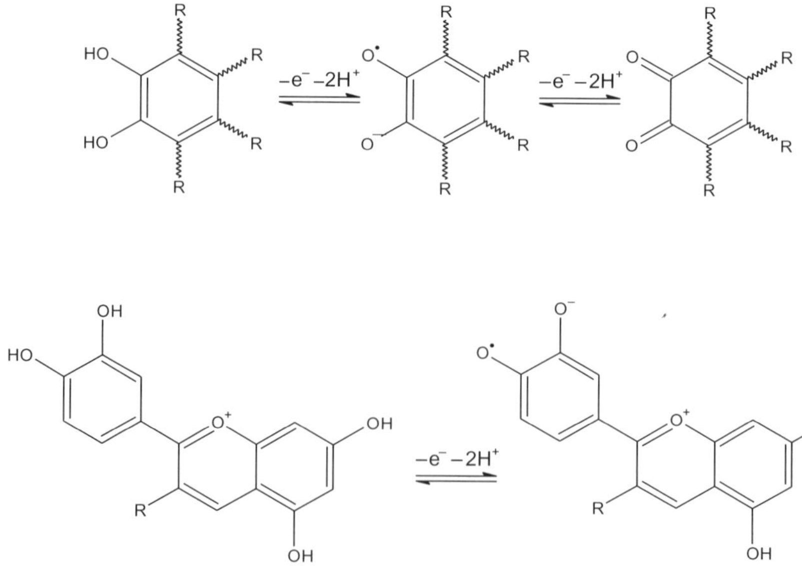

Fig. 3. Figure showing formation of semiquinone radicals in melanin (upper) and anthocyanin (lower).

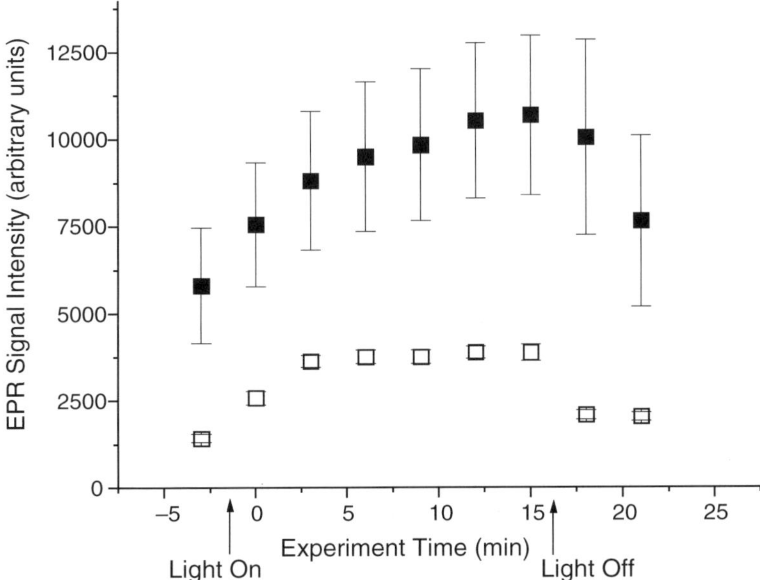

Fig. 4. Time course of light-induced radical production in leaves of *Cornus stolonifera*. Highly pigmented (red) and non-red pigmented leaves were removed and sampled by EPR. Strips of leaf material free of major veins, 1.5 cm by 0.5 cm, were held in a specially-designed polytetrafluorethylene (PTFE) sample holder (Timmins and Davies, 1993) and illuminated with white light from a 200 W tungsten halogen light filtered through a 3 cm layer of water through the 50% cavity grating. EPR spectra were recorded using a Bruker X-band spectrometer operating with 0.5 mT modulation and 10 mW microwave power. The peak to peak 1st derivative lineheight of the light induced species (g = 2.0043, linewidth 0.6 mT) was measured and plotted. Closed symbols represent red leaves (n = 8), open symbols green (n = 6). We have tentatively assigned this species to an anthocyanin-derived phenoxyl or semiquinone radical species, further studies on anthocyanin-deficient mutants are underway to confirm this.

V. CONCLUSIONS: ARE ANTHOCYANINS PLANT MELANINS?

Comparisons of the functional physiology of anthocyanins and melanin reveal that these two major pigment systems play analogous roles in tissues exposed to potentially harmful solar radiation. In addition to their roles in signaling and display, anthocyanins and melanins share the dual physiological capacity to act as sunscreens and to detoxify ROS induced by light. These physiological parallels, in turn, reflect structural similarities between the two pigment systems, as well as the fact that the synthesis of both pigments can be induced by high light intensities. Thus it is clear that highly divergent lines of evolution have evolved–biochemically disparate, but functionally convergent means to dispose of excess light. Of the two pigment systems, melanins have received much more attention. More work is needed to understand the degree to which anthocyanins can be viewed as plant melanins. However, one thing that is abundantly clear is the degree to which anthocyanins and melanins play multiple roles in the physiology of the superficial, light penetrated tissues of both plants and animals.

ACKNOWLEDGMENTS

The authors would like to thank David Lee for his support, encouragement and collaboration. This work was supported by the Andrew W. Mellon Foundation.

REFERENCES

Abuja, P. M., Murkovic, M. and Pfannhauser, W. (1998). Antioxidant and prooxidant activities of elderberry (*Sambucus nigra*) extract in low-density lipoprotein oxidation. *Journal of Agricultural and Food Chemistry* **46**, 4091–4096.

Alscher, R. G., Donahue, J. L. and Cramer, C. L. (1997). Reactive oxygen species and antioxidants: relationships in green cells. *Physiologia Plantarum* **100**, 224–233.

Asada, K. (1994). Mechanisms for scavenging reactive molecules generated in chloroplasts under light stress. *In* "Photoinhibition of Photosynthesis: From Molecular Mechanisms to the Field" (N. R. Baker and J. R. Bowyer, eds), pp. 129–142. Bios Scientific Publishers, Oxford.

Asada, K. (1999). The water cycle in chloroplasts: scavenging of active oxygen and dissipation of excess photons. *Annual Review of Plant Physiology and Plant Molecular Biology* **50**, 601–639.

Azizi, E., Lusky, A., Kushelevsky, A. P. and Schewach-Millet, M. (1988). Skin type, hair color, and freckles are predictors of decreased minimal erythema ultraviolet radiation dose. *Journal of the American Academy of Dermatology* **19**, 32–38.

Baker, N. R. and Hardwick, J. (1973). Biochemical and physiological aspects of leaf development in cocoa (*Theobroma cacao*): I. Development of chlorophyll and photosynthetic activity. *New Phytologist* **72**, 1315–1324.

Balakumar, T., Vincent, V. H. B. and Paliwal, K. (1993). On the interaction of UV-B radiation (280–315 nm) with water stress in crop plants. *Physiologia Plantarum* **87**, 217–222.

Beggs, C. J. and Wellmann, E. (1985). Analysis of light-controlled anthocyanin formation in coleoptiles of *Zea mays* L.: the role of UV-B, blue, red, and far red light. *Photochemistry and Photobiology* **41**, 481–486.

Blarzino, C., Mosca, L., Foppoli, C., Coccia, R., De Marco, C. and Rosei, M. A. (1999). Lipoxygenase/H202-catalyzed oxidation of dihydroxyindoles: Synthesis of melanin pigments and study of their antioxidant properties. *Free Radical Biology and Medicine* **26**, 446–453.

Bongue-Bartelsman, M. and Phillips, D. A. (1995). Nitrogen stress regulates gene expression of enzymes in the flavonoid biosynthetic pathway of tomato. *Plant Physiology and Biochemistry* **33**, 539–546.

Bors, W., Heller, W., Michel, C. and Saran, M. (1990). Flavonoids as antioxidants: determination of radial-scavenging efficiencies. *Methods in Enzymology* **186**, 343–355.

Bors, W., Michel, C. and Saran, M. (1994). Flavonoid antioxidants: rate constants for reactions with oxygen radicals. *Methods in Enzymology* **234**, 20–437.

Brouillard, R. and Dangles, O. (1993). Favonoids and flower color. *In* "The Flavonoids: Advances in Research since 1980" (J. B. Harborne, ed.), pp. 525–538. Chapman and Hall, London.

Bustamante, J., Bredeston, L., Malanga, G. and Mordoh, J. (1993). Role of melanin as a scavenger of active oxygen species. *Pigment Cell Research* **6**, 348–353.

Caldwell, M. M., Robbertecht, R. and Flint, S. D. (1983). Internal filters: prospects for UV-acclimation in higher plants. *Physiologia Plantarum* **58**, 445–450.

Cesarini, J. P. and Msika, P. (1995). Photoprotection from UV-induced pigmentations and melanin introduced in suncreens. *In* "Melanin: Its Role in Human Photoprotection" (L. Zeise, M. R. Chedekel and T. B. Fitzpatrick, eds), pp. 2329–2344. Valdenmar Publishing Company, Overland Park, KS.

Chalker-Scott, L. (1999). Environmental significance of anthocyanins in plant stress responses. *Photochemistry and Photobiology* **70**, 1–9.

Chauhan, N. P., Fatma, T. and Mishra, R. K. (1992). Protection of wheat chloroplasts from lipid peroxidation and loss of photosynthetic pigments by quercetin under strong illumination. *Journal of Plant Physiology* **140**, 409–413.

Christie, P. J., Alfenito, M. R. and Walbot, V. (1994). Impact of low temperature stress on general phenylpropanoid and anthocyanin pathways: enhancement of transcript abundance and anthocyanin pigmentation in maize seedlings. *Planta* **194**, 541–549.

Close, D. C., Beadle, C. L. and Brown, P. H. (2000). Cold-induced photoinhibition affects establishment of *Eucalyptus nitens* (Deane and Maiden) Maden and *Eucalyptus globulus* Labill. *Trees* **15**, 32–41.

Collins, B., Poehler, T. O. and Bryden, W. A. (1995). EPR persistence measurements of UV-induced melanin free-radicals in whole skin. *Photochemistry and Photobiology* **62**, 557–560.

Day, T. A. (1993). Relating UV-B radiation screening effectiveness of foliage to absorbing-compound concentration and anatomical characteristics in a diverse group of plants. *Oecologia* **95**, 542–550.

Deldaldechamp, F., Uhel, C. and Macheiz, J.-J. (1995). Enhancement of anthocyanin synthesis and dihydroflavonol reductase (DFR) activity in response to phosphate deprivation in grape cell suspensions. *Phytochemistry* **40**, 1357–1360.

Demmig-Adams, B. and Adams, W. W. I. (1992). Photoprotection and other response of plants to high light stress. *Annual Review of Plant Physiology and Plant Molecular Biology* **43**, 599–626.

Dixon, R. A., Harrison, M. J. and Lamb, C. J. (1994). Early events in the activation of plant defense responses. *Annual Review of Phytopathology* **32**, 479–501.

Farrington, D., Jr (1964). Man and radiant energy: solar radiation. *In* "Handbook of Physiology: Section 4, Adaptation to the Environment" (D. B. Dill, ed.), pp. 969–987. American Physiological Society, Washington, DC.

Feild, T. S., Holbrook, N. M. and Lee D. W. (2001). Why leaves turn red in autumn. The role of anthocyanins in senescing leaves of red-osier dogwood. *Plant Physiology* **127**, 566–574.

Ferreres, F., Gil, M. I., Castaner, M. and Thomas-Barberan, F. A. (1997). Phenolic metabolites in red pigmented lettuce (*Lactua sativa*). Changes with minimal processing and cold storage. *Journal of Agricultural and Food Chemistry* **45**, 4249–4254.

Fox, H. M. and Vevers, G. (1960). "The Nature of Animal Colors". Sidgwick and Jackson, London.

Foyer, C. H., Lelandais, M. and Kunnert, K. J. (1994). Photoxidative stress in plants. *Physiologia Plantarum* **92**, 696–717.

Gilbert, I. R., Jarvis, P. G. and Smith, H. (2001). Proximity signal and shade avoidance differences between early and late successional trees. *Nature* **411**, 792–795.

Gould, K. S., Kuhn, D. N., Lee, D. W. and Oberbauer, S. F. (1995). Why leaves are sometimes red. *Nature* **378**, 241–242.

Gould, K. S., Markham, K. R., Smith, R. H. and Goris, J. J. (2000). Functional role of anthocyanins in the leaves of *Quintinia serrata* A. Cunn. *Journal of Experimental Botany* **51**, 1107–1115.

Grace, S. C. and Logan, B. A. (2000). Energy dissipation and radical scavenging by the plant phenylpropanoid pathway. *Philosophical Transactions of the Royal Society B.* **355**, 1499–1510.

Grace, S. C., Logan, B. A. and Adams, W. W. I. (1998). Seasonal differences in foliar content of chlorogenic acid, a phenylpropanoid antioxidant, in *Mahonia repens. Plant Cell & Environment* **21**, 513–521.

Hagiwara, A., Miyashita, K., Nakanishi, T., Sano, M., Tamano, S., Kadota, T., Koda, T., Nakamura, M., Imaida, K., Ito, N. and Shirai, T. (2001). Pronounced inhibition by a natural anthocyanin, purple corn color, of 2-amino-1-methyl-6-phenylimidazo[4,5-b]yridne (PhIP) associated colorectal carcinogenesis in male F344 rates pretreated with 1,2-dimethylhydrazine. *Cancer Letters* **171**, 17–25.

Harborne, J. B. (1967). "Comparative Biochemistry of the Flavonoids". Academic Press, New York.

Harborne, J. B. (1988a). The flavonoids: recent advances. *In* "Plant Pigments" (T. W. Goodwin, ed.), pp. 299–343. Academic Press, New York.

Harborne, J. B. (1988b). "The Flavonoids: Advances in Research Since 1980". Chapman and Hall, London.

Harborne, J. B. and Grayer, R. J. (1988). The anthocyanins. *In* "The Flavonoids: Advances in Research since 1980" (J. B. Harborne, ed.), pp. 1–20. Chapman and Hall, London.

Hill, H. Z. (1992). The function of melanin, or 6 blind people examine an elephant. *Bioessays* **14**, 49–56.

Hoch, W. A., Zeldin, E. L. and McGown, B. H. (2001). Physiological significance of anthocyanins during autumnal leaf senescence. *Tree Physiology* **21**, 1–8.

Hrazdina, G., Wagner, G. J. and Siegelman, H. W. (1978). Subcellular localization of enzymes of anthocyanin biosynthesis in protoplasts. *Phytochemistry* **17**, 53–56.

Igarashi, K., Kimura, Y. and Takenaka, A. (2000). Preventive effects of dietary cabbage acylated anthocyanins on paraquat-induced oxidative stress in rats. *Bioscience, Biotechnology and Biochemistry* **64**, 1600–1607.

Jimbow, K., Fitzpatrick, T. B. and Quevedo, J. W. C. (1986). The skin of mammals: Formation, chemical composition and function of melanin pigments. *In* "Biology of the Integument: 2 Vertebrates" (J. Bereiter-Hahn, A. G. Matoltsy and K. S. Richards, eds), pp. 278–292. Springer-Verlag, Berlin.

Juniper, B. E. (1994). Flamboyant flushes: a reinterpretation of non-green flush colors in leaves. *International Dendrology Society Yearbook 1993*, 49–57.

Kaneyuki, T., Noda, Y., Traber, M. G., Mori, A. and Packers, L. (1999). Superoxide anion and hydroxyl radical scavenging activities of vegetable extracts measured using electron spin resonance. *Biochemistry and Molecular Biology International* **47**, 979–989.

Kimura, Y., Araki, Y., Takenaka, A. and Igarashil, K. (1999). Protective effects of dietary nasunin on paraquat-induced oxidative stress in rats. *Bioscience, Biotechnology and Biochemistry* **63**, 799–804.

Kobayashi, N., Nakagawa, A. and Muramatsu, T. (1998). Supranuclear melanin caps reduce ultraviolet induced DNA photoproducts in human epidermis. *Journal of Investigations in Dermatology* **110**, 806–810.

Koes, R. E., Quattrocchio, F. and Mol, J. N. M. (1994). The flavonoid biosynthetic pathway in plants: function and evolution. *Bioessays* **16**, 123–132.

Kollias, N., Sayre, R. M., Zeise, L. and Chedekel, M. R. (1991). Photoprotection by melanin. *Journal of Photochemistry and Photobiology B* **9**, 135–160.

Krol, M., Gray, G. R., Hurry, V. M., Öquist, G., Malek, L. and Huner, N. P. A. (1995). Low-temperature stress and photoperiod affect an increased tolerance to photoinhibition in *Pinus banksiana* seedlings. *Canadian Journal of Botany* **73**, 1119–1127.

Kunz, S., Burkhardt, G. and Becker, H. (1994). Riccionidins A and B, anthocyanidins from the cell walls of the liverwort *Ricciocarpus natans*. *Phytochemistry* **35**, 233–235.

Lee, D. W. and Collins, T. M. (2001). Phylogenetic and ontogenetic influences on the distribution of anthocyanins and betacyanins in leaves of tropical plants. *International Journal of Plant Science* **162**, 1141–1153 .

Lee, D. W., Brammeier, S. and Smith, A. P. (1987). The selective advantages of anthocyanins in developing leaves of mango and cacao. *Biotropica* **19**, 40–49.

Lee, D. W., O'Keefe, J., Holbrook, N. M. and Feild, T. S. (2002). Pigment dynamics and autumn leaf senescence in a New England deciduous forest. Submitted for publication.

Markham, K. R., Gould, K. S., Winefield, C. S., Mitchell, K. A., Bloor, S. J. and Boase, M. R. (2000). Anthocyanic vacuolar inclusions – their nature and significance in flower colouration. *Phytochemistry* **55**, 327–336.

Marrs, K. A., Alfenito, M. R., Lloyd, A. M. and Walbot, V. (1995). A glutathione S-transferase involved in vacuolar transfer encoded by the maize gene *Bronze-2*. *Nature* **375**, 397–400.

Matsuoka, L. Y., Wortsman, J., Haddad, J. G., Kolm, P. and Hollis, B. W. (1991). Racial pigmentation and the cutaneous synthesis of vitamin-D. *Archives of Dermatology* **127**, 536–538.

McClure, J. W. (1975). Physiology and functions of flavonoids. *In* "The Flavonoids" (J. B. Harborne, ed.), pp. 971–1055. Chapman and Hall, London.

Mendez, M., Jones, D. G. and Manetas, Y. (1999). Enhanced UV-B radiation under field conditions increases anthocyanin and reduces the risk of photoinhibition but does not affect growth in the carnivorous plant *Pinguicula vulgaris*. *New Phytologist* **144**, 275–282.

Merzlyak, M. N. and Chivkunova, O. B. (2000). Light-stress-induced pigment changes and evidence for anthocyanin photoprotection in apples. *Journal of Photochemistry and Photobiology B*. **55**, 155–163.

Meyskens, F. L., Farmer, P. and Fruehauf, J. P. (2001). Redox regulation in human melanocytes and melanoma. *Pigment Cell Research* **14**, 148–154.

Mitchell, D. L. (1995). Ultraviolet radiation damage to DNA. *In* "Molecular Biology and Biotechnology" (R. A. Meyers, ed.), pp. 939–943. VCH Publishers, Inc., New York.

Mol, J., Jenkins, G., Schafer, E. and Weiss, D. (1996). Signal perception, transduction and gene expression involved in anthocyanin biosynthesis. *Critical Reviews in Plant Science* **15**, 525–557.

Morison, W. L. (1995). Is melanin a sunscreen? *In* "Melanin: Its Role in Human Photoprotection" (L. Zeise, M. R. Chedekel and T. B. Fitzpatrick, eds), pp. 103–108. Valdenmar Publishing Co., Overland Park, KS.

Neer, R. M. (1975). The evolutionary significance of vitamin D, skin pigment and ultraviolet light. *American Journal of Physical Anthropology* **43**, 409–416.

Neill, S. O. (2002). The functional role of anthocyanins in leaves. Ph.D. thesis, University of Auckland, Auckland, 170 pp.

Neill, S. O. and Gould, K. S. (1999). Optical properties of leaves in relation to anthocyanin concentration and distribution. *Canadian Journal of Botany* **77**, 1777–1782.

Nishio, J. N. (2000). Why are higher plants green? Evolution of the higher plant photosynthetic pigment complement. *Plant Cell and Environment* **23**, 539–548.

Nordlund, J. J. and Boissy, R. E. (2001). The biology of melanocytes. *In* "The Biology of the Skin" (R. K. Freinkel and D. T. Woodley, eds), pp. 113–131. The Parthenon Publishing Group, Pearl River, New York.

Pathak, M. A. and Fanselow, D. L. (1983). Photobiology of melanin pigmentation: dose/response of skin to sunlight and its contents. *Journal of the American Academy of Dermatology* **9**, 724–733.

Pietrini, F. and Massacci, A. (1998). Leaf anthocyanin content changes in *Zea mays* L. grown at low temperature: Significance for the relationship between the quantum yield of PSII and the apparent quantum yield of CO_2 assimilation. *Photosynthesis Research* **58**, 213–219.

Polle, A. (1997). Defense against photooxidative damage in plants. *In* "Oxidative Stress and Molecular Biology of Antioxidant Defenses" (J. G. Scandalios, ed.), pp. 623–666. Cold Spring Harbour Press, Cold Spring Harbour, New York.

Post, A. (1990). Photoprotective pigment as an adaptive strategy in the Antarctic moss *Ceratodon purpureus*. *Polar Biology* **10**, 241–246.

Post, A. and Meret, V. (1992). Photosynthesis, pigments, and chloroplast ultrastructure of an Antarctic liverwort from sun-exposed and shaded sites. *Canadian Journal of Botany* **70**, 2259–2264.

Prota, G. (2000). Melanins, melanogenesis and melanocytes: looking at their functional significance from the chemist's viewpoint. *Pigment Cell Research* **13**, 283–293.

Quevedo, W. C. Jr., Fitzpatrick, T. B., Pathak, M. A. and Jimbow, K. (1975). Role of light in human skin color variation. *American Journal of Physical Anthropology* **43**, 393–408.

Ramirez-Tortosa, C., Andersen, O. M., Gardner, P. T., Morrice, P. C., Wood, S. G., Duthie, S. J., Collins, A. R. and Duthie, G. G. (2001). Anthocyanin-rich extract decreases indices of lipid peroxidation and DNA damage in vitamin E-depleted rats. *Free Radical Biology and Medicine* **31**, 1033–1037.

Riley, P. A. (1997). Melanin. *International Journal of Biochemistry and Cell Biology* **29**, 1235–1239.

Rostand, S. G. (1997). Ultraviolet light may contribute to geographic and racial blood pressure differences. *Hypertension* **30**, 150–156.

Rozanowska, M., Sarna, T., Land, E. J. and Truscott, T. G. (1999). Free radical scavenging properties of melanin interaction of eu- and pheo-melanin models with reducing and oxidising radicals. *Free Radical Biology and Medicine* **26**, 518–525.

Sakihama, Y., Mano, J., Sano, S., Asada, K. and Yamasaki, H. (2000). Reduction of phenoxyl radicals mediated by monodehydroascorbate reductase. *Biochemical and Biophysical Research Communications* **279**, 949–954.

Sharma, V. and Banerji, D. (1981). Enhancement of Hill activity by anthocyanins under both 'white incandescent' and green irradiation (*Rosa damascena* and *Euphorbia pulcherrima*). *Photosynthetica* **15**, 540–542.

Sherwin, H. W. and Farrant, J. M. (1998). Protection mechanisms against excess light in the resurrection plants *Ceratostigma wilmsii* and *Xerophyta viscosa*. *Plant Growth Regulation* **24**, 203–210.

Shirley, B. W. (1996). Flavonoid biosynthesis: 'new functions' for an 'old pathway'. *Trends in Plant Science* **1**, 301–317.

Smillie, R. M. and Hetherington, S. E. (1999). Photoabatement by anthocyanin shields photosynthetic systems from light stress. *Photosynthetica* **36**, 451–463.

Timmins, G. S. and Davies, M. J. (1993) *EPR Newsletter IERC* **5**, 18.

Trull, M. C., Guiltinan, M. J., Lynch, J. P. and Deikman, J. (1997). The response of wild-type and ABA mutant *Arabidopsis thaliana* plants to phosphorus starvation. *Plant Cell and Environment* **20**, 85–92.

Tsuda, T., Ohshima, K., Kawakishi, S. and Osawa, T. (1996). Oxidation products of cyanidin 3-O-beta-D-glucoside with a free radical initiator. *Lipids* **31**, 1259–1263.

Tsuda, T., Horio, F. and Osawa, T. (1998). Dietary cyanidin 3-O-beta-D-glucoside increases *ex vivo* oxidation resistance of serum in rats. *Lipids* **33**, 583–588.

Tsuda, T., Horio, F., Kitoh, J. and Osawa, T. (1999a). Protective effects of dietary cyanidin 3-O-beta-D-glucoside on liver ischemia-reperfusion injury in rats. *Archives of Biochemistry and Biophysics* **368**, 361–366.

Tsuda, T., Horio, F. and Osawa, T. (1999b). Absorption and metabolism of cyanidin 3-O-beta-D-glucoside in rats. *FEBS Letters* **449**, 179–182.

Tsuda, T., Horio, F. and Osawa, T. (2000a). The role of anthocyanins as an antioxidant under oxidative stress in rats. *Biofactors* **13**, 133–139.

Tsuda, T., Kato, Y. and Osawa, T. (2000b). Mechanism for the peroxynitrite scavenging activity by anthocyanins. *Federation of the European Biochemical Societies* **484**, 207–210.

Wheldale, M. (1916). "The Anthocyanin Pigments of Plants". Cambridge University Press, Cambridge.

Wolff, K., Jimbow, K. and Fitzpatrick, T. B. (1974). Experimental pigment donation *in vivo*. *Journal of Ultrastructural Research* **47**, 400–419.

Woodall, G. S. and Stewart, G. R. (1988). Do anthocyanins play a role in UV protection of red juvenile leaves of *Syzygium*? *Journal of Experimental Botany* **49**, 1447–1450.

Yamasaki, H. (1997). A function of colour. *Trends in Plant Science* **2**, 7–8.

Yamasaki, H., Uefugi, H. and Sakihama, Y. (1996). Bleaching of the red anthocyanin induced by superoxide radical. *Archives of Biochemistry and Biophysics* **332**, 183–186.

Anthocyanins in Leaves: Distribution, Phylogeny and Development

DAVID W. LEE[1,2]

[1]*Department of Biological Sciences, Florida International University, Miami, FL 33199, USA*
[2]*Fairchild Tropical Garden, Miami, FL 33156, USA*

ABSTRACT

Red pigments, products of different metabolic pathways, occur in terrestrial plants. The flavonoid pathway contributes the greatest diversity, culminating in the prevalence of anthocyanins in the angiosperms. Anthocyanins are produced in flowers and fruits, and also in vegetative organs, but have been poorly researched in the latter. Anthocyanins are commonly produced in:

1. rapidly expanding leaves of tropical plants;
2. senescing leaves of temperate plants;
3. undersurfaces of floating leaves of aquatic plants;
4. abaxial surfaces of leaves of understory plants; and
5. leaves subjected to various environmental stresses.

The distribution of anthocyanins in leaves, both in presence and in tissue distribution, is influenced by both phylogeny and development. Few species produce anthocyanins in leaf tissues derived from both dermal and ground embryonic tissue. These influences will be important in resolving the ecological roles of anthocyanins in leaves.

Advances in Botanical Research Vol. 37
incorporating Advances in Plant Pathology
ISBN 0-12-005937-1

I. INTRODUCTION

An important context for understanding the functions of anthocyanins in plants is knowledge of their distributions, among organs and in tissues within organs. Certain functions require specific tissue distributions, and tissue distributions may be influenced by evolutionary history and developmental constraints. This article addresses these and other factors influencing anthocyanin distribution. First, it is important to review the history of this research and also to acknowledge that red coloration may be due to the production of other pigments.

A. HISTORICAL BACKGROUND

With the invention of the microscope it was not difficult to observe the distribution of pigments in specific cell layers of plant organs. Anthocyanins are sequestered in vacuoles, comprising the bulk of the volumes of most plant cells. Without any chemical knowledge, this color was viewed as caused by a single class of pigments, forming the 'colored cell sap' in plant organs. Marquart (1835) called the pigment 'anthocyanin'. Observers found anthocyanins in discrete layers within leaves. Morren (1858; Plate 1) observed the pigments in epidermal layers of red cabbage and a native lily. Hassack (1886) used his observations of pigment distributions in leaves to argue for the photoprotective function of these pigments. Later, Stahl (1896; Plate 2) observed pigmentation in both mesophyll and epidermal tissues of leaves of tropical rainforest understory plants. Such descriptions of variations in tissue distributions of anthocyanins in leaves led to some comprehensive surveys. Parkin (1918) reported on the results of a survey of 400 species of plants. Many of these species produced pigments within the mesophyll, but the report omitted distributions in specific cell layers and did not list the species surveyed. An extensive survey by Gertz (1906; described by Wheldale, 1916) showed that anthocyanins were more commonly produced in certain families and suggested there might be some systematic relationships between the presence of anthocyanins among different families.

B. RED PIGMENTATION IN TERRESTRIAL PLANTS

Although the large majority of red-violet colors observed in plants are due to anthocyanins, many exceptions occur, and other pigments have evolved (Fig. 1). Most of these pigments are produced by the flavonoid pathway, but other metabolic pathways produce red pigments, including the terpenoid pathway. A variety of structural modifications shift absorbances toward longer wavelengths in xanthophylls (Moss and Weedon, 1976); some absorb at wavelengths above 500 nm and produce red colors, such as lycopene and rhodoxanthin (Fig. 1A). All pigments vary in occurrence among major plant groups, but the anthocyanins are by far the most important in the angiosperms.

Some bryophytes produce red shoots, particularly in response to low temperatures; these are particularly common in New Zealand forests. Sphagnorubin (Fig. 1E) is produced in *Sphagnum* and a few other bryophytes, and is a flavonoid aglycone perhaps derived from 3-deoxy anthocyanidin (Fig. 1D), bound to cell walls and distinct from anthocyanins (Markham, 1988; Mues, 2000). Similar red flavonoid aglycones, riccionidins A and B (Fig. 1H), have been observed in the cell walls of the liverwort *Ricciocarpos natans* and in three other genera. Such a compound may have been observed by Post and Vesk (1992) in the cold and high irradiance responses of an Antarctic liverwort, *Cephaloziella exiliflora*. Some bryophytes may redden in response to low temperatures by producing xanthophylls. Post (1990) observed the induction of violaxanthin in the photoprotection of an Antarctic moss, *Ceratodon purpureus*. In general, anthocyanins may be unusual in the bryophytes (Markham, 1988; Geiger *et al.*, 1997; Mues, 2000) but may be common in certain areas (as New Zealand, Kevin Gould, personal communication) and certainly deserve more study among these plants.

Some pteridophytes produce red leaves, particularly during expansion. *Selaginella erythropus*, popular in cultivation, produces red undersurfaces from the accumulation of an unknown pigment. Although anthocyanins have been reported from a variety of ferns (Soeder, 1985), these may not be true anthocyanins, but 3-deoxy anthocyanidins (Fig. 1D). These are flavonoid pigments, but are derived from flavanones rather than the flavones and 3-hydroxy flavanones as anthocyanins are (Stafford, 1994). Thus, they are biosynthetically slightly different than anthocyanins. Such pigments have been observed in a variety of ferns (Harborne, 1966; Markham, 1988).

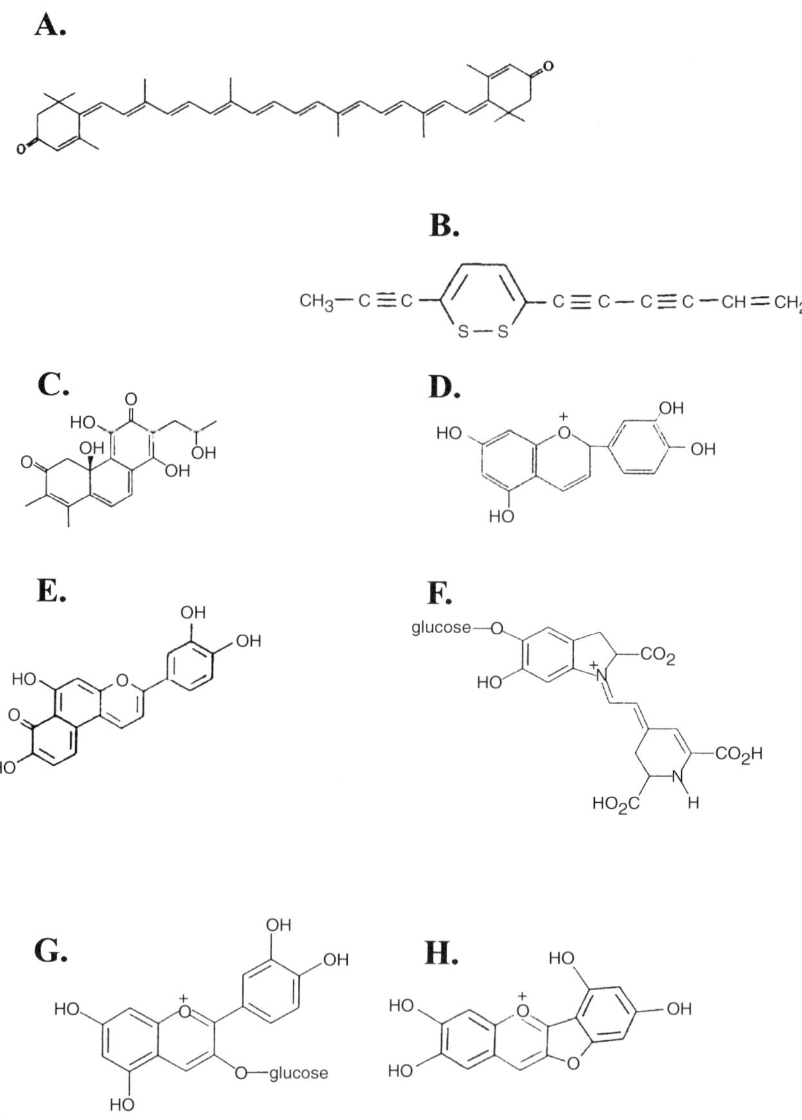

Fig. 1. Red pigment molecules that have been elucidated from various terrestrial plants. (A) Rhodoxanthin, a xanthophyll (Goodwin, 1976); (B) Thiarubrin-A, a polyacetylene from several taxa in the Asteraceae (Rodriguez *et al.*, 1985); (C) Coleone-E, a triterpenoid from *Coleus* sp. (Menthaceae; Thompson, 1976)); (D) 3-deoxy anthocyanidin, common in ferns and in a few angiosperms (Harborne, 1966); (E) Sphagnorubin, sporadically isolated in members of the Bryophyta (Markham, 1988); (F) Betanin, a nitrogenous pigment found exclusively in members of the Centrospermae, the core Caryophyllales (Stafford, 1994); (G) Cyanidin-3-glucoside, the most commonly isolated anthocyanin from leaves of angiosperms (Harborne, 1976); and (H) Riccionidin-A, seen in the cell walls of a few liverworts (Mues, 2000).

Gymnosperms produce red or brown-red pigmentation, during development or when induced under low temperatures. Goodwin (1976) reported the production of rhodoxanthin (Fig. 1A) in winter needles of *Cryptomeria*. In a survey of tropical taxa that included five gymnosperms, Lee and Collins (2001) observed reddish-brown expanding leaves in three species, all of these due to the accumulation of pigments in plastids. These were probably xanthophylls and certainly not flavonoids. Yet, some gymnosperm species do produce anthocyanins, particularly in response to cold. In *Pinus banksiana* seedlings, low temperatures induced the production of anthocyanin (cyanidin-3-glucoside; Fig. 1G). This helped increase tolerance to photoinhibition (Krol *et al.*, 1995). Niemann (1988) reported a few examples of foliar anthocyanin production, primarily in the Podocarpaceae.

Anthocyanins occur in angiosperms in considerable diversity (Gianassi, 1988), but the principal type present in most leaves, early in development and during senescence, is cyanidin-3-glucoside (Fig. 1G; Harborne, 1976). Closer inspection using contemporary analytical techniques is beginning to reveal a greater diversity, particularly as different glycones and modifications of the cyanidin structure. Although red-violet coloration in flowering plants is generally due to anthocyanin accumulation, there are notable exceptions.

The most significant is the production of the nitrogenous betacyanin pigments in the Centrospermae, or core Caryophyllales. Species in this group of families produce flavonoids, but not anthocyanins, and betacyanins (as betanin; Fig. 1F) are responsible for reddish colors in flowers and vegetative organs (Stafford, 1994). There are other exceptions. In the Gesneriaceae reddish colors may be due to 3-deoxy anthocyanidin (Fig. 1D); reported in ferns as previously mentioned (Harborne, 1966). Reddish colors may also be produced by xanthophylls, as rhodoxanthin (Fig. 1A) in *Potamogeton* (Goodwin, 1976). Ida *et al.* (1995) observed the reddening of *Buxus sempervirens* leaves to be due to an unusual red carotenoid, anhydroeschscholtzxanthin. Xanthophyll pigments were responsible for reddish senescing leaves in *Colubrina elliptica* (Lee and Collins, 2001). Unusual pigments may cause the reddening of vegetative tissues in some plants. Thiarubrin-A, a polyacetylene compound (Fig. 1B) in *Aspilia* and other genera in the Asteraceae, produces the red color in an oil that is primarily in roots but also in lower branches and leaves (Rodriguez *et al.*, 1985). Thompson (1976) mentioned other unusual molecules present as red pigments in plants. Coleone (Fig. 1C) is a triterpenoid pigment produced in the leaves of *Coleus*. Fuerstione is a similar molecule present in leaf glands of *Fuerstia Africana*, also in the Menthaceae.

II. ANTHOCYANINS IN ANGIOSPERMS

Anthocyanins are produced in virtually all vegetative organs in flowering plants. They are produced in exposed roots, as the adventitious roots of *Ficus benghalensis* (unpublished observation). They are frequently produced in young shoots, including buds. They are produced in leaves, often in petioles as well as leaf blades. In leaves anthocyanins are present in:

1. mesophyll and abaxial epidermis of forest understory plants;
2. abaxial epidermis of floating aquatic plants as *Nymphaea*;
3. during leaf senescence;
4. during leaf expansion; and
5. tissues produced as a response to stress, as from nutrient deficiency, extreme temperatures, high irradiance and disease (Chalker-Scott, 1999 and this volume).

With the exception of the Centrospermae, anthocyanins are thought to be constitutively produced in all angiosperms. However, in a survey of 399 tropical species Lee and Collins (2001) observed 79 (19.8%) with no visible evidence of anthocyanin production.

Anthocyanin distribution within leaf tissues and among taxonomic groups is relevant to understanding their function in leaves. If these pigments have a physiological function in leaves, then we would expect their presence and distributions to vary with the environmental conditions that would be mitigated by this function, and for them to be absent in conditions without any selective advantage for pigmentation. Anthocyanins may not only differ in their presence but also their distributions in leaf tissues. The general assumption has been that anthocyanins are produced in epidermal layers of leaves, perhaps providing defense against damage by UV radiation. Although the documentation from earlier research is poor, such early surveys showed that anthocyanins were produced in the mesophyll of many plants (Wheldale, 1916; Parkin, 1918). Recent results by Gould and Quinn (1999) and Lee and Collins (2001) reveal considerable variation in the tissue distribution among different species.

A. ENVIRONMENTAL FACTORS

Although we lack good quantitative comparative data, anthocyanin production in leaves appears to be associated with certain environmental factors. Within species, anthocyanin production, both in development and senescence, may be enhanced by exposure to high irradiance. Anthocyanin production also appears to be enhanced by low temperatures (perhaps in association with high irradiance). This may occur during leaf senescence, or annually in evergreen species whose leaves last several years. Leaf undersurface coloration is more frequent in species of forest understory in both temperate and tropical regions. These are environments of extreme shade, with brief exposures to the high irradiances of sunflecks. Finally, a variety of stresses during the leaf life span may induce the synthesis of anthocyanins. Abilities of plants to produce anthocyanins may be the result of the natural selection of traits for the production of these molecules, or the plasticity to produce them in response to environmental stimuli.

In addition to ecology (or habitat partitioning), two other factors may influence anthocyanin distribution:

1. phylogenetic influences, or inertia; and
2. developmental constraints.

These constraints on anthocyanin production may obscure the ecological relationships.

B. PHYLOGENY

There is a growing literature on the analysis of potential phylogenetic influences on character distribution in organisms (Felsenstein, 1985; Harvey and Pagel, 1991; and Miles and Dunham, 1993). The critical element in such analyses is a robust phylogeny. The molecular phylogenies routinely produced in plant systematics provide the frameworks for such analyses, and the potential benefits are beginning to be appreciated (Lord *et al.*, 1995; Ackerly and Donoghue, 1998; Ackerly and Reich, 1999).

Here I discuss some implications for such a large-scale analysis, very recently published (Lee and Collins, 2001). In this analysis of 399 woody tropical taxa, characters of anthocyanin tissue distribution during leaf expansion and senescence were mapped on a large-scale three-gene molecular phylogeny of the angiosperms (Soltis *et al.*, 1999). We determined the presence of red-violet pigmentation in vacuoles of leaf cell layers or features: adaxial epidermis, hypodermis (if present), palisade mesophyll, spongy mesophyll, bundle sheath cells only, abaxial epidermis, and trichomes. Additional structure within the family level was provided from other research, and the phylogeny was pruned to include families represented by the 399 species.

As an illustration of character distribution in one portion of the phylogeny, the Malvaceae is illustrated (Fig. 2). This large family is an amalgamation of four traditional families of primarily tropical woody plants: Malvaceae, Sterculiaceae, Bombaceae and Tiliaceae (APG, 1998). Additional structure in this tree was provided by recent molecular

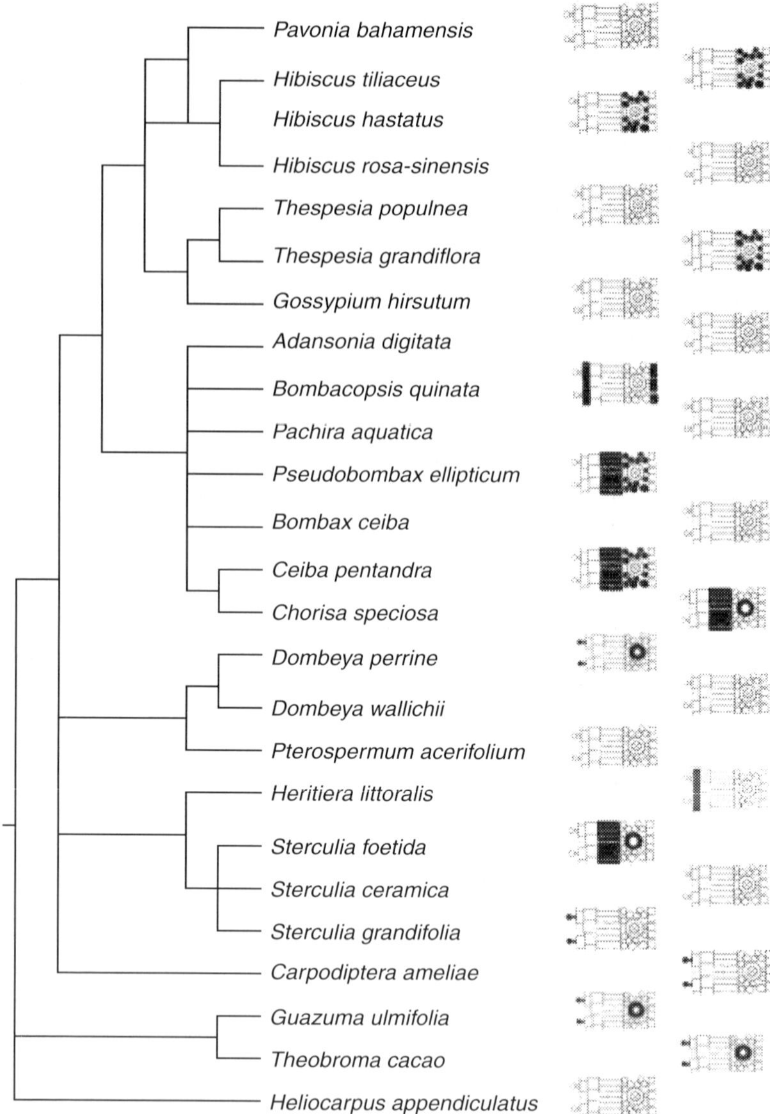

Fig. 2. The distribution of anthocyanins in leaf tissues of expanding leaves of species within the Malvaceae (APG, 1998) surveyed by Lee and Collins (2001). The cartoons depict the locations of anthocyanins in vacuoles of cells of the adaxial epidermis, hypodermis, palisade parenchyma, spongy mesophyll, bundle sheath cells, abaxial epidermis, trichomes.

research (Alverson *et al.*, 1998; Whitlock *et al.*, 1999; David Baum, personal communication), and the distribution of taxa among the families explains the establishment of a larger Malvaceae. In this figure the presence of anthocyanins in different cell layers is illustrated for individual species. Fourteen of 25 species produced anthocyanins in leaf tissues in development, and these were produced in six different tissue combinations. Five of these species were unusual among the larger sample in producing anthocyanins in bundle sheath cells; there was only one other example (*Strongylodon macrobotrys*, Fabaceae) in the entire sample. This tissue combination was found in species of three of the four clades in this tree; only the taxa formerly in the Malvaceae *sensu stricto* had no such character.

These results visually suggest considerable variation in anthocyanin distribution within a single large family, particularly in presence and absence of production as well as in tissue distribution. However, the presence of anthocyanins in bundle sheath cells was almost unique to this family.

Distribution of these characters for leaf expansion within the large data set can also be compared between families by selecting only those with five or more species in the sample, and then presenting the mean for the family (Fig. 3). Here the tissue distributions are summarized for the germ layers from which they are derived, either dermal or ground tissue (or in a few cases, both), for 32 families. The clades represent the major lineages in the angiosperms. We see large differences among different families. For instance, almost all members of the Myrtaceae produced anthocyanins (mostly in the palisade mesophyll), and all members of the Combretaceae and Lythraceae. Thus the Myrtales has members with pronounced production of anthocyanins during development. On the other hand, few members of the Asteraceae, Araliaceae, Verbenaceae and Solanaceae produced anthocyanins in development or senescence. Thus the percentage production was lower in the Asteridae, particularly the Euasteridae II (APG, 1998). However, The Bignoniaceae had a fairly high percentage of members producing anthocyanins, particularly in dermal tissue.

The distribution of anthocyanins during senescence was quite different than during development. In the phylogeny of the Malvaceae (Fig. 2) only three species of the 25 species produced anthocyanins during senescence: *Carpodiptera ameliae*, *Gossypium hirsutum* and *Dombeya perrine*. Only seven families in the larger scale phylogeny (Fig. 3) produced 40% or more species with anthocyanins during leaf senescence: Euphorbiaceae, Malpighiaceae, Polygonaceae, Bignoniaceae, Flacourtiaceae, Combretaceae and Lythraceae. Only in the last three families did red leaf senescence appear in the large majority of the species sampled. Most families in this largely tropical survey did not develop any anthocyanins during leaf senescence.

In this sample of tropical taxa the percentage of taxa producing anthocyanins during senescence was much lower than during development, 13.5% versus 44.9%. This contrasts strongly with a second sample of 90 woody species from mixed deciduous forests in Central Massachusetts

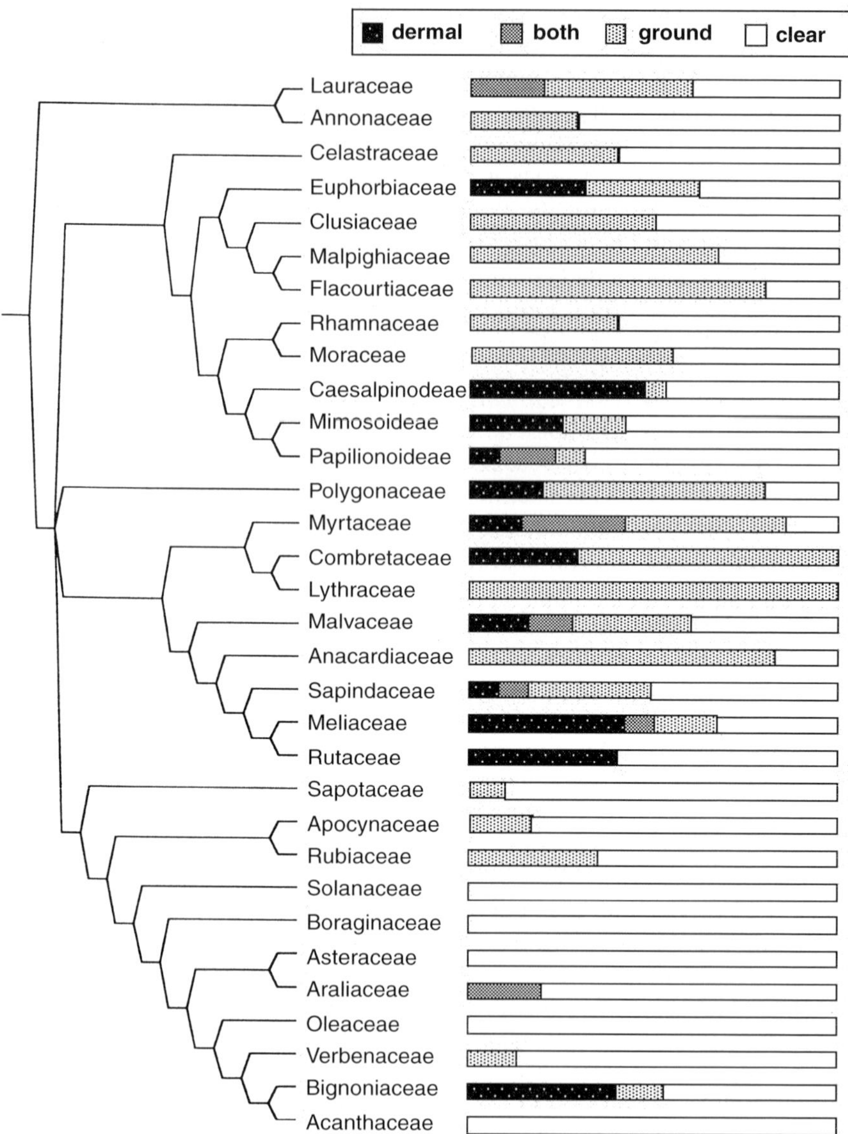

Fig. 3. Means of tissue presences in developing leaves of various angiosperm families for which there was a minimal sample of five species, in the survey of Lee and Collins (2001). Bar graphs indicate the frequency of presence of anthocyanins in dermal tissues (adaxial and abaxial epidermis and trichomes), ground tissues (hypodermis, palisade parenchyma, bundle sheath and spongy mesophyll cells), both, or absent from all tissues.

(Lee *et al.*, 2002). The large majority (62, or 70%) of these species produced anthocyanins during leaf senescence. All but two of them produced anthocyanins in palisade mesophyll cells, and 11 species in both epidermis and palisade mesophyll cells. Certain families were particularly well-represented with anthocyanic taxa: Caprifoliaceae, Cornaceae and Rosaceae. Other families were poorly represented, as the Betulaceae (Lee *et al*., 2002).

Mapping tissue distributions of anthocyanins on a phylogeny certainly indicates that the distributions are not random. The real strength of this approach is that it is amenable to hypothesis testing by statistical analysis. Are these distributions influenced by phylogeny, suggestive of phylogenetic 'inertia' (Maddison and Slatkin, 1991)? This question was tested on the phylogeny of 399 species in the following way (Lee and Collins, 2001). We generated both random character distributions and random resolutions of the polytomies on the study tree. We then compared the numbers of steps for characters of the resolutions of the study tree with the number of steps with the randomly shuffled character states. We also calculated retention indices; when RI = 1, phylogenetic inertia is maximized. We found that the number of steps differed significantly in four characters during development: epidermis (RI = 0.14), palisade parenchyma (RI = 0.25), spongy mesophyll (RI = 0.19) and lower epidermis (RI = 0.16). Steps differed during senescence only for palisade parenchyma (RI = 0.24). These differences were weak, yet highly significant, and suggest that these five characters were influenced by evolution. The relative ease of loss of expression due to mutation probably partially obscures this inertia. Thus, in this sample of tropical plants phylogeny played some role in the distribution of anthocyanins in leaf tissues.

C. DEVELOPMENT

A variety of factors independent of phylogeny may constrain the distributions of anthocyanins in leaf tissues. Consider the number of tissue layers or structures in which anthocyanins can be produced: adaxial epidermis, hypodermis (if present), palisade parenchyma, spongy mesophyll, bundle sheath cells, adaxial epidermis, and trichomes (if present). The first and last two layers/structures are dermal in embryological origin, and the rest originate from ground tissue (Westhoff, 1998; Sinha, 1999). Anthocyanins could theoretically be produced in these different tissues singly or in multiple combinations, for a total 127 combinations (126 if the absence of anthocyanins is subtracted), or 2^7.

Lee and Collins (2001) found that a small fraction of this total occurred in their sample, both during leaf expansion and senescence (Table I). During expansions a mere nine tissue combinations accounted for 88.4% of all leaves producing anthocyanins. During senescence only seven tissue combinations accounted for 96.3% of all leaves. Ground tissue (particularly palisade parenchyma) was the prevalent site of

TABLE I

Distribution of anthocyanins in tissues of developing and senescent leaves in a survey of 399 tropical woody plants (Lee and Collins, 2001). These combinations are the most frequent among those observed in the survey.

Tissue type[a]									
UEP	HYP	PAL	SPM	VBU	LEP	TRC	Tissue origin	% in development	% in senescence
–	–	*[b]	*	–	–	–	ground	26.3[c]	18.5
–	–	*	–	–	–	–	ground	19.0	50.0
–	–	–	*	–	–	–	ground	10.6	0.0
–	–	*	–	*	–	–	ground	3.4	0.0
–	–	–	–	*	–	–	ground	3.4	0.0
*	–	–	–	–	*	–	dermal	14.5	1.9
*	–	–	–	–	–	–	dermal	9.5	7.4
*	*	*	–	–	–	–	both	0.0	7.4
*	–	*	–	–	–	–	both	0.6	7.4
*	*	–	–	–	–	–	both	1.1	3.7
Total								88.4 (9)	96.3 (7)

[a] UEP = adaxial epidermis; HYP = hypodermis, if present; PAL = palisade parenchyma; SPM = spongy mesophyll; VBU = parenchyma adjacent to vascular bundles; LEP = abaxial epidermis; TRC = trichomes or scales.

[b] * indicates presence in cell vacuoles in this tissue layer.

[c] Values are in % of taxa with anthocyanins, and numbers in parentheses are the numbers of tissue combinations accounting for the percentages.

anthocyanin accumulation during expansion (62.7%) and senescence (68.5%).

Dermal tissue accumulated anthocyanins in much lower percentages both during expansion (24.0%) and senescence (9.3%). The most striking result is the low percentages of leaves producing anthocyanins on tissues of both germ lines, only 6.7% out of the species producing anthocyanins in development, 18.5% in senescence.

There are seven possible tissue combinations of dermal origin ($2^3 - 1$) and 15 combinations of ground origin ($2^4 - 1$). We observed six of those combinations of dermal origin, and 13 of those combinations of ground origin. Surprisingly, there are 105 possible combinations of the tissues of both origins, and we observed only nine combinations. A conservative test of the significance of this deviation was performed by comparing the observed character combinations with those of 100 random simulations. Anthocyanin expression in both dermal and ground tissues was less in our data than in the simulations ($P<0.01$).

Thus, there are developmental constraints on the production of anthocyanins in leaves, against certain tissue combinations. Most of the loci controlling anthocyanin expression in plant tissues affect dermal tissue (Dooner *et al.*, 1991; Mol *et al.*, 1996), and those concerning ground (mesophyll) tissue are fewer and less studied (Kubo *et al.*, 1999). In maize, virtually any tissue combination in any organ can be reproduced experimentally (Virginia Walbot, personal communication). Yet, relatively few tissue combinations, and therefore more limited systems of genetic control, are commonly encountered in nature, revealed in this survey.

D. ADDITIONAL INFLUENCES

Other factors may affect anthocyanin expression in leaves. Production of anthocyanins in other plant parts, particularly flowers and fruits, may be correlated with anthocyanins in leaves. Since these organs develop from leaf-like segments, such correlations would not be surprising (Fineblum and Rausher, 1997). Lee and Collins (2001) found that 38% of the 127 species without anthocyanins in other plant parts produced anthocyanins in the leaf blade in expansion and/or senescence. In the majority of species producing anthocyanins in other organs, 53% produced anthocyanins in leaves.

The expression of anthocyanins in leaves during development may be ancestral to expression during senescence (or *visa versa*), and the patterns were similar or identical in those 35 species. In three heteroblastic vines (*Ficus pumila*, *Marcgravia rectiflora* and *Macfadyena unguis-cati*) the patterns of anthocyanin production were identical at all stages. These are further minor constraints on anthocyanin expression in leaves. In other heteroblastic species, patterns of anthocyanin distribution may differ among leaves of heteroblastic stages. Patterns differ in leaf stages of *Hedera helix* (see Hackett in this volume), as they also do in different stages of the New Zealand tree, *Pseudopanax crassifolius* (Gould, 1993).

III. SUMMARY

Leaves of flowering plants are generally distinguished by the frequent production of anthocyanins at different stages in the leaf life span. Such production is rare in bryophytes, pteridophytes and seed plants, although other red pigments may be produced. In flowering plants, anthocyanin distribution in leaves is clearly influenced by:

1. ecological pressures;
2. phylogeny; and
3. developmental constraints.

The influence of development and phylogeny complicate the analysis of the relationship between ecology and chemistry. There are even larger more difficult questions. Within families and similar functional types, as trees of a genus growing in tropical deciduous forests (many of the plants analyzed by Lee and Collins, 2001), some plants produce anthocyanins and others do not. If anthocyanins confer some physiological/selective advantage, how do the species lacking anthocyanins compensate? Certain families do not produce anthocyanins in leaves. If there is an advantage, how do they compensate? Finally, a series of families in the Caryophyllales produce betacyanins and not anthocyanins. Some of the species in the Nyctaginaceae and other families produce betacyanins during leaf expansion, perhaps analogous to anthocyanin production in other families.

Do these molecules, chemically quite distinct but with similar absorbance and antioxidant activity, perform a similar physiological function?

Although we have not moved very far to attack these problems, what is now encouraging is that we have arrived at a level of understanding in different disciplines to begin working cooperatively to solve them. This volume illustrates the future directions of that research.

ACKNOWLEDGMENTS

Several colleagues provided technical information and critiqued the manuscript: Tim Collins, Kelsey Downum, Tim O'Shea and Kevin Gould. David Baum helped with the phylogeny of the Malvaceae. The idea for this article (and the symposium) was catalysed by a visit to New Zealand to work with Kevin Gould, supported by a Royal Society of New Zealand Marsden Grant.

REFERENCES

Ackerly, D. D. and Donoghue, M. J. (1998). Leaf size, sapling allometry, and Corner's rules: phylogeny and correlated evolution in maples (*Acer*). *American Naturalist* **152**, 767–791.

Ackerly, D. and Reich, P. B. (1999). Convergence andcorrelations among leaf size and function in plants: A comparative test using independent contrasts. *American Journal of Botany* **86**, 1272–1281.

Alverson, W. S., Karol, K. G., Baum, D. A., Chase, M. W., Swensen, S. M., McCourt, R. and Sytsma, K. J. (1998). Circumscription of the Malvales and correlations to other Rosidae: evidence from *rbc*L sequence data. *American Journal of Botany* **85**, 876–887.

The Angiosperm Phylogeny Group (APG) (1998). An ordinal classification for the families of flowering plants. *Annals of the Missouri Botanical Garden* **85**, 531–553.

Chalker-Scott, L. (1999). Environmental significance of anthocyanins in plant stress responses. *Photochemistry and Photobiology* **70**, 1–9.

Dooner, H. K., Robbins, T. P. and Jorgensen, R. A. (1991). Genetic and developmental control of anthocyanin biosynthesis. *Annual Review of Genetics* **25**, 173–199.

Felsenstein, J. (1985). Phylogenies and the comparative method. *American Naturalist* **125**, 1–15.

Fineblum, W. L. and Rausher, M. E. (1997). Do floral pigmentation genes also influence resistance to enemies? The *W* locus in *Ipomoea purpurea*. *Ecology* **78**, 1646–1654.

Geiger, H., Seeger, T., Zinsmeister, H. D. and Frahm, J.-P. (1997). The occurrence of flavonoids in arthrodontous mosses – an account of the present knowledge. *Journal of the Hattori Botanical Laboratory* **83**, 273–308.

Gertz, O. (1906). "Studier öfver anthocyan". *Akademisk Afhandling*, Lund.

Gianassi, D. (1988). Flavonoids and evolution in the dicotyledons, *In* "The Flavonoids" (J. B. Harborne, ed.), pp. 479–504. Chapman and Hall, London.

Goodwin, T. W. (1976). Distribution of carotenoids. *In* "Chemistry and Biochemistry of Plant Pigments, Volume 1" (T. W. Goodwin, ed.), pp. 225–261. Academic Press, London.

Gould, K. R. (1993). Leaf heteroblasty in *Pseudopanax crassifolius*. Functional significance of leaf morphologyand anatomy. *Annals of Botany* **71**, 61–70.

Gould, K. R. and Quinn, B. D. (1999). Do anthyocyanins protect leaves of New Zealand native species from UV-B? *New Zealand Journal of Botany* **37**, 15–18.

Harborne, J. B. (1966). Comparative biochemistry of flavonoids – II. 3-desoxyanthocyanins morphology and their systematic distribution in ferns and gesneriads. *Phytochemistry* **5**, 589–600.

Harborne, J. B. (1976). Functions of flavonoids in plants. *In* "Chemistry and Biochemistry of Plant Pigments, Volume 1" (T. W. Goodwin, ed.), pp. 736–779. Academic Press, London.

Harvey, P. and Pagel, M. (1991). "The Comparative Method in Evolutionary Biology". Oxford University Press, Oxford.

Hassack, C. (1886). Untersuchungen über den anatomischen Bau bunter Laublätter, nebst einigen Bemerkungen, betreffend die physiologische Bedeutung der Buntfärbung derselben. *Botanisches Centrablatt fuer Deutschland* **28**, 84–387.

Heller, W. and Forkmann, G. (1988). Biosynthesis. *In* "The Flavonoids" (J. B. Harborne, ed.), pp. 399–425. Chapman and Hall, London.

Holton, T. A. and Cornish, E. C. (1995). Genetics and biochemistry of anthocyanin synthesis. *Plant Cell* **7**, 1071–1083.

Ida, K., Masamoto, K., Maoka, T., Fuhiwara, Y., Takeda, S. and Hasegawa, E. (1995). The leaves of the common box, *Buxus sempervirens* (Buxaceae), become red as the level of a red carotenoid, anhydroeschscholtzxanthin, increases. *Journal of Plant Research* **108**, 369–376.

Krol, M., Gray, G. R., Hurry, V. M., Öquist, G., Malek, L. and Huner, N. P. A. (1995). Low-temperature stress and photoperiod affect an increased tolerance to photoinhibition in *Pinus banksiana* seedlings. *Canadian Journal of Botany* **73**, 1119–1127.

Kubo, H., Peeters, A. J. M., Aarts, M. G. M., Pereira, A. and Koornneef, M. (1999). ANTHOCYANINLESS2, a homeobox gene affecting anthocyanin distribution and root development in *Arabidopsis*. *Plant Cell* **11**, 1217–1226.

Lee, D. W. and Collins, T. M. (2001). Phylogenetic and ontogenetic influences on the distribution of anthocyanins and betacyanins in leaves of tropical plants. *International Journal of Plant Science* **162**, 1141–1153.

Lee, D. W., O'Keefe, J., Holbrook, N. M. and Feild, T. S. (2002). Pigment dynamics and autumn leaf senescence in a New England deciduous forest. Submitted for publication.

Lord, J. L., Westoby, M. and Leishman, M. (1995). Seed size and phylogeny in six temperate floras: constraints, niche conservatism, and adaptation. *American Naturalist* **146**, 349–364.

Maddison, W. P. and Slatkin, M. (1991). Null models for the number of evolutionary steps in a character on a phylogenetic tree. *Evolution* **45**, 1184–1197.

Markham, K. R. (1988). Distribution of flavonoids in the lower plants and its evolutionary significance. *In* "The Flavonoids" (J. B. Harborne, ed.), pp. 427–468. Chapman and Hall, London.

Marquart, L. C. (1835). "Die Farben der Blüthen, Eine Chemisch-physiologische Abhandlung". Habichte, Bonn.

Miles, D. B. and Dunham, A. E. (1993). Historical perspectives in ecology and evolutionary biology: the use of phylogenetic comparative analyses. *Annual Review of Ecology and Systematics* **24**, 587–619.

Mol, J., Jenkins, G., Schafer, E. and Weiss, D. (1996). Signal perception, transduction and gene expression involved in anthocyanin biosynthesis. *Critical Reviews in Plant Science* **15**, 525–557.

Morren, E. (1858). "Dissertation sur les feuilles vertes et Colorées". C. Annoot-Braeckman, Gand, Belgium.

Moss, G. P. and Weedon, B. C. L. (1976). Chemistry of carotenoids. *In* "Chemistry and Biochemistry of Plant Pigments, Volume 1" (T. W. Goodwin, ed.), pp. 149–224. Academic Press, London.

Mues, R. (2000). Chemical constituents and biochemistry. *In* "Bryophyte Biology" (A. J. Shaw and B. Goffinet, eds), pp. 150–181. Cambridge University Press, Cambridge.

Niemann, G. J. (1988). Distribution and evolution of the flavonoids in gymnosperms. In "The Flavonoids" (J. B. Harborne, ed.), pp. 469–478. Chapman and Hall, London.

Parkin, J. (1918). On the localisation of anthocyanin (red-cell sap) in foliage leaves. *Report of the British Association for the Advancement of Science* **1903**, p. 862. J. Murray, London.

Post, A. (1990). Photoprotective pigment as an adaptive strategy in the Antarctic moss *Ceratodon purpureus*. *Polar Biology* **10**, 241–243.

Post, A. and Vesk, M. (1992). Photosynthesis, pigments, and chloroplast ultrastructure of an Antarctic liverwort from sun-exposed and shaded sites. *Canadian Journal of Botany* **70**, 2259–2264.
Rodriguez, E., Aregullin, M., Nishida, T., Uehara, S., Wrangham, R., Abramowski, Z., Finlayson, A. and Towers, G. H. N. (1985). Thiarubrine A, a bioactive constituent of *Aspilia* (Asteraceae) consumed by wild chimpanzees. *Experientia* **41**, 419–420.
Sinha, N. (1999). Leaf development in angiosperms. *Annual Review of Plant Physiology and Plant Molecular Biology* **50**, 419–448.
Soeder, R. W. (1985). Fern constituents: Including occurrence, chemotaxonomy and physiological activity. *Botanical Review* **51**, 442–536.
Soltis, P. S., Soltis, D. E. and Chase, M. W. (1999). Angiosperm phylogeny inferred from multiple genes as a tool for comparative biology. *Nature* **402**, 402–404.
Stafford, H. A. (1994). Anthocyanins and betalains: evolution of the mutually exclusive pathways. *Plant Science* **101**, 91–98.
Stahl, E. (1896). Ueber bunte Laublätter. *Annals of the Botanical Garden Buitenzorg* **13**, 137–216.
Thompson, R. H. (1976). Miscellaneous pigments. *In* "Chemistry and Biochemistry of Plant Pigments, Volume 1" (T. W. Goodwin, ed.), pp. 597–623. Academic Press, London.
Westhoff, P. (1998). "Molecular Plant Development". Oxford University Press, Oxford.
Wheldale, M. (1916). "The Anthocyanin Pigments of Plants". Cambridge University Press, Cambridge.
Whitlock, B. A., Nyffeler, R., Bayer, C. and Baum, D. A. (1999). Phylogeny of the core Malvales: evidence from NDHF sequence data. *American Journal of Botany* **86**, 1474–1486.

The Final Steps in Anthocyanin Formation: A Story of Modification and Sequestration

CHRISTOPHER WINEFIELD[1,2]

[1]*New Zealand Institute for Crop and Food Research Ltd, Private Bag 11600, Palmerston North, New Zealand*
[2]*Current address: Biology Department, University of York, P.O. Box 373, Heslington Road, York YO10 5YW, UK*

ABSTRACT

Over the last few years a committed effort has lead to the elucidation of much of the biosynthetic and regulatory pathways leading to the formation of anthocyanins in plant tissues. Through a combination of biochemical and reverse genetic approaches, the genes encoding the major biosynthetic steps have been cloned. Indeed, the highly visible phenotype of the anthocyanins has facilitated the identification of mutants and their corresponding genes for practically the entire pathway. However while our knowledge about the formation of these compounds is substantial, there are still a number of interesting facets to the pathway that remain to be elucidated. In particular recent efforts by a number of groups have sought to clarify the mechanisms that are responsible for the final stages of biosynthesis and sequestration of these pigments. Identification of the specific genes/enzymes involved in these terminal steps has proved difficult as these reactions involve the action of proteins belonging to large multigene families that commonly share overlapping functions within the plant. The glycosyltransferases, glutathione S-transferases and membrane transporters that have known functions in these terminal stages of anthocyanin biosynthesis, in particular have been shown to belong to large gene families in which closely related members exhibit overlapping activities. This has made elucidation of these terminal modification and transportation steps difficult to study, as useful mutants that would point to specific function within this pathway have been elusive. In this chapter, recent efforts that address the final stages in anthocyanin biosynthesis, the interplay between the physical environment of the vacuole and anthocyanin function, and

Advances in Botanical Research Vol. 37
incorporating Advances in Plant Pathology
ISBN 0-12-005937-1

recent work investigating aspects of sequestration will be discussed in the context that these processes have on the functionality of anthocyanins *in vivo*.

I. INTRODUCTION

The vivid phenotypes imparted to plant tissues that accumulate anthocyanins has ensured that the genetics and more recently the biochemistry of their formation has been the focus of various investigative efforts for over 200 years. Indeed it is the visibility of the phenotype that has, in conjunction with modern molecular approaches to both biochemistry and molecular biology, allowed the near complete elucidation of this rather complex multienzyme pathway. Mutations and mutants for practically the entire biosynthetic pathway have been identified and in many instances it has been the characterisation of these mutants that has allowed the identification and cloning of the gene(s) encoding specific biosynthetic activities. This work has been extended to the identification of various regulatory factors that co-ordinately regulate the pathway at the transcriptional level. The reader is encouraged to seek out several excellent reviews that have been recently published that deal in depth with the biosynthesis and regulation of this pathway (Shirley, 1996; Mol *et al.*, 1998, 1999; Winkel-Shirley, 2001).

As one looks toward the terminal steps of anthocyanin biosynthesis it is evident that only recently has there been an increasing appreciation of the interplay between the actions of the enzymes responsible for the modifications and placement of anthocyanins within the cell, and anthocyanin chemistry and function. The recent resurgence in interest in this area has been hampered by a combination of factors of which arguably the most important has been the overlapping functionality between the proteins involved in these processes. This ability to compensate for a loss of the primary activity has ensured that for a majority of species it has been difficult to identify gene specific mutations that target these final steps.

The recent and ongoing release of genome data for plant species like arabidopsis, maize and rice has greatly facilitated the investigation of large and complex multigene families. This information has allowed researchers the use of bioinformatic approaches to identify complete gene families and then pursue the functional analysis of their corresponding members (McGonigle *et al.*, 2000; Keegstra and Raikhel, 2001; Li *et al.*, 2001). Of the growing number of articles published in this area that are of interest to workers in the area of anthocyanin research, probably the most significant to be recently published describes the identification of the entire glycosyltransferase multigene family from arabidopsis (Li *et al.*, 2001). This family consists of 107 members that exhibit a wide range of activities. Phylogenetic analysis of this family reveals a strong linkage between activity and sequence relationships (Li *et al.*, 2001). This work also highlights that for a given substrate, a number of overlapping activities may be active in any given tissue. This highlights the importance of both sound biochemical investigation coupled with genetic evidence to identify those enzymes specifically involved with anthocyanin biosynthesis at these latter stages in the pathway.

Work to characterise the glutathione S-transferase (GST) gene family in maize and soybean has shown that these enzymes also form a large and complex multigene family (McGonigle *et al.*, 2000). Work with those GSTs involved in the transport of anthocyanins has clearly highlighted the difficult task of identification of the specific enzymes involved. As will be discussed later, *Bz2* and *An9* from maize and petunia respectively, have been shown to encode the specific and functionally equivalent GSTs involved in anthocyanin transport, yet they only share 20% amino acid identity (Alfenito *et al.*, 1998). This clearly shows that identification of specific GSTs involved in anthocyanin transport is a difficult task without the availability of genetically defined mutants.

Even with our growing understanding of the mechanisms by which anthocyanins are synthesised, it is apparent that we need to have a wider appreciation of not only the biosynthesis but also the chemical and environmental impacts on the anthocyanin molecule and how these affect function/colour *in vivo* (Figueiredo *et al.*, 1996). As will be discussed later this is only one of several considerations that have to be taken into account when addressing the reproduction or introduction of new colours *in vitro* and *in vivo* respectively (Figueiredo *et al.*, 1996). It is now becoming evident that the physical environment of the vacuole or other cellular locations where anthocyanins are deposited has practically the same influence on the final colour and function of the accumulating anthocyanins as do the steps that lead to the production of these molecules.

Not only, therefore, are the steps physically involved with the synthesis of the anthocyanins important, but so too is the physical organisation of the pathway and how it functions to transport these products to the vacuole or other compartment within or outside the cell space. Interesting work has begun to look at the potential for using this pathway to answer questions as to the existence of large multienzyme complexes (Burbulis and Winkel-Shirley, 1999). This work is suggestive of the formation of a large multienzyme complex that is localised to the endoplasmic reticulum (ER) and anchored to the cytoplasmic face of this membrane by various cytochrome P450 enzymes (P450) associated with the pathway (Burbulis and Winkel-Shirley, 1999; Winkel-Shirley this volume).

Work to identify the specific transportation mechanism involved in the movement of anthocyanins to their final localisation has produced a number of candidate transporters (Lu *et al.*, 1998; Rea *et al.*, 1998; Debeaujon *et al.*, 2001). There is strong evidence for transport across the tonoplast membrane being mediated by vanadate sensitive multidrug resistance-associated protein (MRP) related ABC like transporters (Marrs, 1996). On the other there is also evidence for the transport of anthocyanins from the proposed site of synthesis on the ER, through an ER vesicle mediated transport system and delivery that is subsequently achieved through fusion of these vesicles with the tonoplast (Grotewold, 2001).

Once delivered to the vacuole there has been clearly shown to be both intramolecular and intermolecular interactions between individual anthocyanins and with co-localised flavonoid compounds. These interactions allow a stabilisation and enhancement of colour through a process termed copigmentation (Brouillard and Dangles, 1994). Recent work investigating the occurrence of anthocyanin aggregates within the vacuole has suggested the potential presence of other interactions within the vacuole that might further serve to stabilise and sequester the anthocyanins within the vacuole (Markham *et al.*, 2000a).

Within this chapter a number of the issues outlined above will be expanded to review work that has sought to investigate the varying aspects of the final stages of anthocyanin production, anthocyanin transport and the interplay between the chemistry of the anthocyanin molecule and the surrounding physical environment. Finally, work investigating the occurrence of anthocyanin aggregates will be reviewed in the context of the role that these products may play within the vacuole.

II. STRUCTURE VERSUS FUNCTION

The biosynthesis of anthocyanins can be considered a dual pathway with two major intertwined processes at work. The first is the production of a 'functional' anthocyanin complement in the tissues that are programmed for the production and accumulation of these pigments. The second is the invocation of a system whose purpose is the inactivation and removal of toxic materials from the cell cytoplasm. Indeed it is clear that upon close examination, a convergent evolutionary process has developed by which the plant has utilised differing pathways to produce and cope with the production of potentially toxic materials that have proven highly beneficial to plant survival.

As I will discuss later, aspects of the biosynthesis of anthocyanins strongly resemble the pathways invoked for the detoxification of xenobiotics (Coleman *et al.*, 1997). In particular the hydroxylation and glycosylation reactions as well as the involvement of glutathione S-transferases (GSTs) in the transportation of anthocyanins to the vacuole, are characteristic of this pathway, and are strikingly conserved with the phase II and III reaction scheme of xenobiotic detoxification (Coleman *et al.*, 1997). These modifications, however, do not serve only as a means of correctly locating the anthocyanins to their final resting place. They also have profound affects on the final colour and function attributed to a particular anthocyanin structure.

A. HYDROXYLATION AND ITS IMPACT ON THE ANTHOCYANIN CHROMOPHORE

Probably the set of reactions that have the greatest impact on anthocyanin function and colour are catalysed by the action of three P450 monoxygenase enzymes in particular (Winkel-Shirley, 2001). The first, flavone-3-hydroxylase (F3H) acts to add a hydroxyl group to the 3 position of the A ring of the developing anthocyanin structure (Fig. 1). Mutations to the F3H gene result in the formation of colourless/white tissues, as in the majority of plants, the subsequent enzyme in the pathway, dihydroflavonol reductase (DFR), is incapable of using the accumulating flavanones as a substrate. It is also significant that this hydroxyl group is required for the initial glycosylation of the anthocyanin carried out by UDP-glucose dependant glycosyltransferases, such as those encoded by the *Bz2* gene from maize (Ralston *et al.*, 1988). Addition of the glucose to the 3-position of the A ring of the anthocyanin requires the presence of this hydroxyl group. The bronze phenotype of the *bz1* mutation is indicative that this reaction is also required for proper sequestration/compartmentalisation of the accumulating anthocyanin (Jones and Vogt, 2000; Vogt and Jones, 2000). In

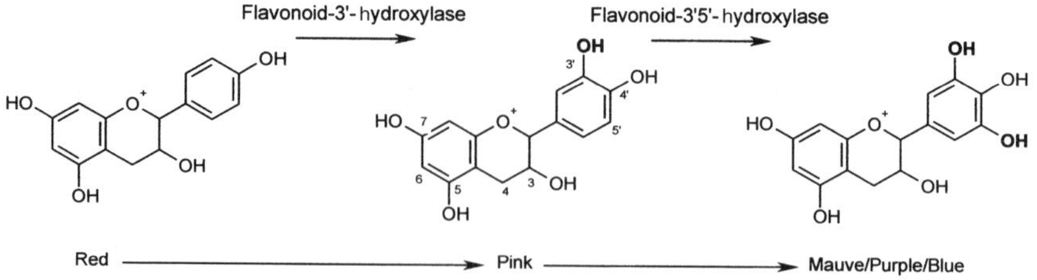

Fig. 1. The structures of the three major anthocyanin aglycones, pelargonidin, cyanidin and delphinidin. Sequential hydroxylation by the cytP450 enzymes flavonoid-3′-hydroxylase and flavonoid-3′5′-hydroxylase add hydroxyl groups to the 3′ and 5′ positions of the B-ring respectively (highlighted in bold type). This hydroxylation shifts the absorbance spectra of the molecule such that the colour of the anthocyanin changes from red toward blue.

addition it has been shown the resulting 3-glycoside is a specific substrate of other glycosyltransferases that act at subsequent positions around the molecule (Yamazaki *et al.*, 1999). These combined issues indicate that even if DFR were capable of utilizing the accumulating flavanones there are issues surrounding the sequestration and stability of the resultant compounds.

The hydroxylation reactions catalysed, at several points throughout the pathway, by two other P450 enzymes, flavonoid-3′-hydroxylase (F3′H) and flavonoid-3′5′-hydroxylase (F3′5′H), are arguably the most important in the formation of the plethora of different colours produced by these compounds in plants (Mol *et al.*, 1998, 1999; Winkel-Shirley, 2001). This fact has made these enzymes a prime target for biotechnological approaches to manipulate plant colour (Mol *et al.*, 1999). The addition of hydroxyl groups to the 3′ and 5′ positions of the B ring (Fig. 1) enables a bathochromic shift in colour from red through to mauve and blue. This quite dramatic shift in absorbance that is in the order of a few nanometres is caused through a shift in the electron density surrounding the chromophore (Brouillard and Dangles, 1994).

B. COPIGMENTATION

While the alteration of the hydroxylation state has a major impact on the absorbance spectra of anthocyanin molecules, subsequent additions of glucose, galactose, acyl and methyl moieties along with interactions with flavonoid molecules, has an equally dramatic impact on colour and function through a process termed copigmentation (Brouillard and Dangles, 1994; Figueiredo *et al.*, 1996; Bloor, 1997). While those genes encoding the hydroxylases have been the focus for biotechnology approaches, aimed at altering the anthocyanin complement in a given species, those genes encoding proteins involved in copigmentation activities have received somewhat less attention.

Work investigating methods for controlling the copigmentation of anthocyanins *in vivo* has been spurred on by initial transgenic studies that have aimed to produce blue pigmentation through addition of the F3′5′H and subsequent production of the tri-hydroxylated anthocyanin delphinidin, in plants such as carnation (Mol *et al.*, 1998, 1999).These experiments revealed that both the degree of copigmentation, as well as factors such as vacuolar pH, have a major impact on the final colour in any given cell, irrespective of the hydroxylation state of the anthocyanin. While the genes encoding the enzymes involved in the production of both flavonols and flavanones have now been cloned and are relatively conserved between species (Holton *et al.*, 1993; Holton and Cornish, 1995; Akashi *et al.*, 1999; Martenss and Forkmann, 1999), those genes encoding the transferases involved with the addition of various moieties around the molecule have proven to be somewhat more difficult to identify conclusively. As described earlier, while it is relatively straightforward to identify large multienzyme families capable of these reactions, due to the overlapping substrate specificities common to these enzymes it is difficult to assign those specifically acting on the anthocyanins.

1. Glycosylation as a Method for Copigmentation

As will be discussed later, glycosylation of the anthocyanin at the 3 position of the A ring is extremely important for the correct targeting of the anthocyanin to the vacuole or indeed other locations within the cell (Ralston *et al.*, 1988). If one looks at isolated anthocyanins from a given plant, they are commonly extensively substituted (Brouillard and Dangles, 1994). Additions can be made to the glycosyl group added to the 3 position or at various points around the molecule. It is clear that these additions act to aid in the stabilisation of the molecule in aqueous environments such as those found in the vacuole, and to cause significant bathochromic shifts in the molecule's absorbance spectra (Figueiredo *et al.*, 1996). Studies investigating the pigments present in blue flowers such as blue marguerite daisy and blue agapanthus serve to show the complex nature of these additions and the interactions that take place with equally decorated flavonoid copigment partners within the vacuole (Bloor, 1999; Bloor and Falshaw, 2000). In each of these cases the copigment complex causes a stabilisation of the pigment and a large shift in absorbance to allow blue colour formation both *in vivo* and *in vitro* (Bloor, 1999; Bloor and Falshaw, 2000). Agapanthus floral pigments and those isolated from the blue flowers of *Evolvulus pilosus* cv. Blue Daze and from the blue-purple flowers of *Eichhornia crassipes*, exhibit remarkable stability as well as a large bathochromic shift in absorbance

(Figueiredo *et al.*, 1996). It is proposed that these additions at flanking positions on the anthocyanin molecule (e.g. 3′ and 7 positions see Fig. 1) can allow extensive self-association with similarly decorated anthocyanins and their respective copigment partners (Bloor, 1997). Data presented from the analysis of the pigments isolated from *Ceanothus* (Californian lilac) are strongly suggestive that in this situation the acyl groups within the extensive additions at the 3′ and 7 positions of the delphinidin molecule can associate with the central (pyrylium) ring of the anthocyanin thereby affording increased stabilisation and a strong bathochromic shift in absorbance due to the inhibition of hydration of the central ring and a resulting shift in electron densities surrounding the chromophore (Figueiredo *et al.*, 1996; Bloor, 1997).

2. Copigmentation with the Flavonoids

The example above clearly shows the importance of glycosylation and other additional activities, such as acyltransferases, for both the stabilisation and the copigmentation of the anthocyanins. However of equal importance are the well-recognised intermolecular interactions afforded by the colourless flavonoids. Experiments that have manipulated the levels of flavonol synthase (FLS) for example, clearly show the importance of flavonols in the production of mauve/blue pigments from delphinidin based anthocyanins produced in lisianthus flowers. In the flowers of lisianthus it was possible to shift the colour of the accumulating anthocyanins toward red by reducing the production of flavonols through the down-regulation of FLS using antisense technologies (Karen Nielsen, personal communication).

C. ENVIRONMENTAL CONSIDERATIONS

As we have discussed, the chemistry of the anthocyanin molecule is particularly important for its stability and function. With this in mind it is probably not surprising to find that the environment into which the anthocyanin is placed also has an impact on these issues. For any given anthocyanin molecule there is an equilibrium between a number of ionic forms, the ratio of each determined in part by the structure and in part by the environmental conditions into which the anthocyanin is placed. Typically anthocyanins placed into weakly acidic (pH 2–4) aqueous solutions will be colourless or weakly coloured due to a displacement of the hydration equilibrium toward the formation of the hemiacetal or chalcone forms (Figueiredo *et al.*, 1996). Increasing degrees of substitution or association with flavonoids, allowing varying degrees of copigmentation, act to ameliorate this shift leading to retention of colour. *In vivo* the pH of the compartment (typically the large central vacuole) into which the anthocyanins are placed will determine the ratios of flavillium ion and quinonoidal base forms of the anthocyanins. This will in turn influence the absorbance spectra (hence colour) of the anthocyanins accumulated in this compartment. Strongly acidic solutions ensure the predominance of the flavillium ion form and in these cases, irrespective of the alterations to structure mentioned earlier, the anthocyanins form a strong red colour in solution. In order for blue colour formation this equilibrium has to favour the formation of the quinonoidal base (Figueiredo *et al.*, 1996). This fact has made the molecular engineering of blue colour formation so difficult over recent years (Mol *et al.*, 1998). Attempts are currently under way to investigate methods for altering vacuolar pH and results appear to be promising (Mol *et al.*, 1998). In certain plants such as *Ceanothus*, the extensive substitutions already discussed offer, through extensive intramolecular copigmentation, another potential means to form blue colouration in 'non-permissive' pH backgrounds.

III. BIOSYNTHESIS VERSUS DETOXIFICATION: SIMILAR PATHWAYS TO THE SAME END?

As mentioned earlier in this chapter the similarities between the biosynthesis of anthocyanins and those schemes put forward for the detoxification of xenobiotics are quite striking. This insightful and compelling idea put forward by Kathleen Marrs and co-workers has underpinned our recent understanding of the processes at work in the latter stages of anthocyanin production and sequestration to the vacuole (Marrs, 1996).

In order to gain an appreciation of this similarity it is necessary to summarise the main biosynthetic steps leading to anthocyanin production. The synthesis of the anthocyanins begins with the action of phenylalanine ammonia lyase (PAL) and involves the action of over ten separate proteins before the final products enter the vacuole (Fig. 2). In the initial phase of synthesis phenylalanine is converted to cinnamic acid, then to ρ-coumaric acid and 4-coumaroyl-CoA by the actions of cinnimate-4-hydroxylase (C4H) and 4-coumyral CoA ligase (4CL) respectively. Chalcone synthase acts at this point to condense the 4-coumaroyl-CoA backbone with three malonyl-CoA subunits to form the first committed step in flavonoid and anthocyanin biosynthesis, the chalcones. The open ring structure of the chalcones can then either close spontaneously or through the action of chalcone isomerase to form the flavanones. It is from this point onward that the pathway shares a commonality between the synthesis of a specific secondary metabolite and detoxification of toxic compounds.

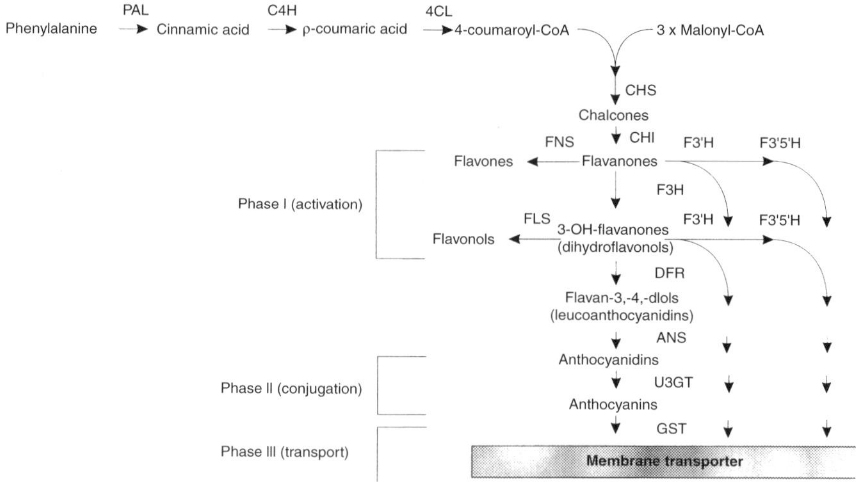

Fig. 2. A stylised representation of the major steps of anthocyanin biosynthesis. The parts of the pathway that show similarity to the three phases of xenobiotic detoxification are marked. The enzymes listed are as follows: PAL – phenylalanine ammonia lyase, C4H – cinnamate-4-hydroxylase, 4CL – 4-coumyral CoA ligase, CHS – chalcone synthase, CHI – chalcone isomerase, FNS – flavone synthase, F3′H – flavonoid-3′-hydroxylase, F3′5′H – flavonoid-3′5′-hydroxylase, F3H – flavone-3-hydroxylase, FLS – flavonol synthase, DFR – dihydroflavonol reductase, ANS – anthocyanidin synthase, U3GT – UDP-dependant 3-O-glycosyltransferase, GST – glutathione S-transferase.

A. ANTHOCYANIN SYNTHESIS VERSUS XENOBIOTIC DETOXIFICATION

Generally the scheme for detoxification of xenobiotics can be broken down into the following three phases (Coleman *et al.*, 1997). An initial phase (phase I) that is characterised by a number of activation steps generally carried out through the action of a range of cytochrome P450 enzymes. These activated molecules are then targets of a range of transferase enzymes that act to add glucose, malonate or glutathione moieties to form water-soluble conjugates (phase II). The final stage (phase III) is the removal of the inactivated, water-soluble conjugates from the cytosol by membrane-located transport proteins that initiate the compartmentalisation and processing part of the detoxification process (Fig. 2; Coleman *et al.*, 1997). Application of this scheme to the anthocyanin pathway, as outlined in the previous section, highlights the striking similarity between the two pathways, one leading to the formation of specific secondary metabolites, the other resulting in the removal of toxic compounds from the cytoplasm.

The reactions carried out by the three main P450 enzymes, F3H, F3′H, and F3′5′H can be considered to carrying out the phase I activation reactions. These enzymes add hydroxyl groups to carbons at positions

3 on the A ring and 3′ and 5′ on the B-ring respectively. The F3′H and F3′5′H enzymes are capable of acting on both the flavanones and the dihydroflavonols (Holton *et al.*, 1993; Holton and Cornish, 1995; Brugliera *et al.*, 1999). The subsequent actions of dihydroflavonol reductase (DFR) and anthocyanidin synthase (ANS; also known as leucoanthocyanidin dioxygenase (LDOX)), lead to the formation of the 3-OH-anthocyanidins that are the substrates for glycosylation activities that act to add a glucose group to the 3-hydroxyl on the A-ring. Thus this and subsequent glycosyl, acyl and methyltransferases can be considered to be carrying out the equivalent of the phase II detoxification reactions.

The importance of the substitution reactions carried out by the transferase enzymes, is highlighted through mutations in these genes and the profound affect this has on the ability of the cell to remove these compounds from the cytoplasm. Of the various mutant phenotypes reported to be associated with the transferase activities (both glycosylation and glutathionation), it has been the *Bz1* and *Bz2* genes that have been interpreted in having a role in transportation (Ralston *et al.*, 1988; Marrs *et al.*, 1995; Marrs, 1996). The *Bz1* allele in maize encodes an UDP dependant glycosyltransferase that conjugates an UDP-glucose moiety to the 3-hydroxyl on the A-ring of anthocyaninidins. Without this addition subsequent interaction with the glutathione S-transferase specifically associated with anthocyanin sequestration (encoded by the *Bz2* allele) is prevented, leading to the accumulation of these products within the cytoplasm (Ralston *et al.*, 1988; Marrs *et al.*, 1995; Marrs and Walbot, 1997). In addition to interference with transportation processes, mutations in this allele interfere with substitution reactions at other positions around the molecule of the anthocyanin. From studies characterising glycosyltransferase activities against the 5-position on the anthocyanin A-ring, a 3 glycosylated anthocyanin substrate appears to be necessary (Yamazaki *et al.*, 1999). Lack of glycosylation also has consequences for the stability of the anthocyanin within the cytoplasm. In the case of both the *bz1* and the *bz2* mutations, oxidation reactions are proposed to take place within the cytoplasm leading to the production of a bronze like colour (Marrs *et al.*, 1995; Marrs, 1996). Despite the oxidation reactions it is also likely, as described earlier, that the anthocyanins will be subject to nucleophillic attack of the central pyrylium ring by water, forcing the formation of the relatively colourless chalcone form of the anthocyanin (Figueiredo *et al.*, 1996).

Investigation of the second ‘transportation’ mutant, *bz2* from maize, revealed the involvement of a glutathione S-transferase in the transportation of anthocyanins. Initial investigations of the possible identity of the mutated allele lead to the observation of the similarity of these latter steps in anthocyanin biosynthesis and xenobiotic detoxification processes. Utilising recombinant proteins it was possible to show that the gene did indeed encode a GST belonging to class III of plant GSTs. Those plant GSTs characterised to date most closely align to the theta class of GSTs. However within this classification, plant GSTs can be further sub-classified into class I, II or III groupings based on amino acid identity and conservation of inton:exon placement (Marrs, 1996; Edwards *et al.*, 2000). Initial work to characterise BZ2 revealed a protein that could utilise the generic GST substrate 1 chloro-2,4-dinitrobenzene (CDNB) and that BZ2 utilised cyanidin 3-glucoside an estimated 15 times more effectively than the aglycone form. Although initial reports indicated that conjugation of cyanidin with glutathione was part of the BZ2 action, recent reports indicate that in fact GST's involved with anthocyanin sequestration from both maize (BZ2) and petunia (AN9) act as anthocyanin carriers and that gluthationation is not required (Alfenito *et al.*, 1998; Mueller *et al.*, 2000). Further studies comparing related class III GSTs from maize and BZ2 have shown that mutations of the critical residues involved with glutathione binding to these GSTs do not affect BZ2's ability to transport anthocyanins. This data strongly suggests that glutathionation is not a prerequisite of anthocyanin transportation and that BZ2 acts as a carrier molecule targeting the anthocyanin to the transporters on the tonoplast membrane (Virginia Walbot, personal communication).

IV. TRANSPORTATION

Evidence gathered from the study of various anthocyanin biosynthetic mutants has clearly shown the importance for the correct intracellular localisation of anthocyanins. Localisation of anthocyanins within the cytoplasm, as is seen in the *bz1* and *bz2* mutants of maize, leads to a bronze type colour caused by the oxidation of the accumulated anthocyanins within the cytoplasm (Marrs *et al.*, 1995). In other plants these blockages may lead to the production of tan or cream/white tissues (Virginia Walbot, personal communication). However in addition to the requirement for sequestration of anthocyanins for function there is also a requirement to remove these and other similar compounds (products of secondary metabolism or xenobiotic detoxification) from the cytoplasm as they represent potentially toxic compounds that could interfere with the normal functioning of important metabolic processes within the cytoplasm.

The involvement of GSTs in the biosynthesis of anthocyanins and the similarities between this pathway and that employed for xenobiotic detoxification led to the speculation of the involvement of multidrug resistance-associated proteins (MRP) like ATP-binding cassette (ABC) like transporters to shuttle the glycosylated anthocyanins from the cytoplasm to the vacuole (Marrs *et al.*, 1995; Marrs, 1996; Rea *et al.*, 1998; Rea, 1999). Experiments in which protoplasts derived from maize aleurone cells were treated with vanadate, a potent inhibitor of transporters that form an acylphosphate during their catalytic cycles, led to a phenocopy of the *Bz2* phenotype (Marrs *et al.*, 1995). This result strongly suggested that a member of the MRP subfamily of ABC transporters (i.e. a glutathione pump) was involved in the final stages of anthocyanin sequestration into the vacuole. To date, however, the gene that encodes this activity has not been cloned. The lack of identifiable mutants for this genetic loci tends to suggest that there may be overlapping functionality of several transporters of this class that can compensate for mutations in any one of the corresponding sub-family (Edwards *et al.*, 2000). A wide distribution for GS-X pump mediated vacuolar anthocyanin uptake in plants is also suggested by the facility of both AtMRP1 and AtMRP2 from arabidopsis, for the high efficiency transport of cyanidin-3-glycoside-glutathione conjugates in recombinant yeast cells that lack MRP like transporter activity (Lu *et al.*, 1997; Lu *et al.*, 1998; Rea *et al.*, 1998).

Recently reports describing the transport of flavonoid glucuronides and other derivatives of phenylpropanoid biosynthesis, indicate that a range of ABC-like transporters are involved and that they have the potential for the transportation of a wide range of related compounds, both endogenously produced and those encountered in the environment (Klien *et al.*, 1996, 2000, 2001; Rea *et al.*, 1998; Walczak and Dean, 2000; Debeaujon *et al.*, 2001). One of these transporters isolated from arabidopsis, encoded by TT12 has recently been cloned and shown to belong to transporters belonging to the multidrug and toxic compound extrusion (MATE) family of transporters (Debeaujon *et al.*, 2001). This family of transporters recently proposed by Brown and co-workers exhibits little sequence similarity to the vanadate-sensitive ABC transporters proposed to be involved in anthocyanin transport (Brown *et al.*, 1999). The *tt12* gene product is proposed to be involved in the transport of flavonoid glycosides in the seed of this plant (Debeaujon *et al.*, 2001).

The existence of multiple transporters that appear to be involved in the transportation of these compounds supports the view that there are overlapping functionalities within the plant to cope with the wide range of compounds that the plant either produces or encounters on a day-to-day basis.

V. THE VACUOLE – INTRA VERSUS INTERCOMPARTMENTAL SEQUESTRATION

It has been generally believed that there is little transformation of anthocyanins once they are transported to the vacuole. The site of action of enzyme activities that may further modify the anthocyanin is unclear at this time. Those reports of enzymes such as the anthocyanin 5-aromatic acyltransferase isolated from *Gentiana triflora* (Fujiwara *et al.*, 1997) and the 5-glycosyltransferase isolated from *Perilla frutescens* (Yamazaki *et al.*, 1999), suggest that these enzymes are localised in the cytoplasm. Additionally, although it has been shown that it is most likely a glutathione conjugate of anthocyanin that is transported across the tonoplast membrane, no glutathione conjugates have been reported to have been isolated from plants. One explanation for this is that upon reaching the vacuole the anthocyanin conjugate is metabolised to remove and recycle the conjugate (Marrs *et al.*, 1995). However, this hypothesis is challenged by the fact it appears that both BZ2 and AN9 act solely as carriers and that glutathionation is not involved in the transportation process (Alfenito *et al.*, 1998; Mueller *et al.*, 2000).

The view of the vacuole as a large storage and/or dumping ground for a wide range of metabolites and xenobiotic compounds is gradually changing as it becomes evident that there are a wide range of essential biosynthetic activities that take place within this compartment. With this in mind it is interesting to note that it is still generally accepted that secondary metabolites and xenobiotics that are placed in the vacuole are free in solution. This view is somewhat incongruous with the fact that we consider that these compounds are toxic and that their removal from the cytoplasm is essential for continued cytoplasmic function. The presence of these compounds, free in vacuolar solution, could potentially lead to a similar situation to that which the cell has striven to avoid in the cytoplasm (i.e. interference with essential metabolic function). Observations of anthocyanin aggregates within the vacuoles of a wide range of species, and reports of alternative sites for flavonoid localisation within the cell is suggestive that a potential method for intracompartmental sequestration of anthocyanins and other secondary metabolites possibly exists.

In terms of anthocyanins, their accumulation within the cytoplasm of certain organs does not in itself appear to be toxic. This is presumably due to a number of reasons, including the ‘deactivating’ substitutions of the base molecule (Coleman *et al.*, 1997). However the accumulation of these compounds within the cytoplasm, in mutants such as *bz1* and *bz2*, must be dealt with in some manner to prevent damage to cytoplasmic components. Recent reports investigating the accumulation of flavonoids in the petals of lisianthus (*Eustoma grandiflorum*), *Lathyrus chrysanthus* and *Dianthus caryophyllus* flowers highlights that in these cases at least, the vacuole may not be the only compartment for the sequestration of these compounds (Markham *et al.*, 2000b; Markham *et al.*, 2001). In these plants it was shown that significant proportions of the flavonoids produced are localised to components in either the cell wall or cytoplasm (Markham *et al.*, 2001). Indeed in the case of yellow *Lathyrus* flowers the pigmentation derives entirely from either cell wall or membrane bound flavonoids. These pigments generally do not contribute to colouration of a given tissue directly. However in this case their contribution to colour derives from binding large amounts of flavonoid to some component of either the cell wall or some unidentified cytoplasmic component. This results in a strong bathochromic shift that allows a yellow colouration of the cells (Markham *et al.*, 2001). Preliminary experiments investigating the possibilities for binding partners for these flavonoids suggested that there might be specific associations with a particular type of protein within the cytoplasm. This suggests that there may be a specific set of 'sequestration' proteins within the cytoplasm to assist in coping with the inappropriate localisation of these potentially toxic compounds within the cytoplasm.

By extension, a similar mechanism may well be at work within the vacuole. To this end a number of observations have lent some weight to the thought that there may be some sort of intracompartmental mechanism for flavonoid/anthocyanin sequestration, that not only assists in the prevention of interference with essential vacuolar processes but may also contribute significantly to the overall function of these compounds *in-planta.*

A. ANTHOCYANIC VACUOLAR INCLUSIONS – PHYSICAL COMPARTMENTATION AND COPIGMENTATION?

Over many years there have been observations of anthocyanin aggregations within the pigment accumulating cells of a wide range of species. These have been described variously as blue spherules in epidermal rose petals (Yasuda, 1974), intravacuolar spherical bodies in *Polygonum cuspudatum* seedlings (Kubo *et al.*, 1995), ball like structures and crystals in stock, *Matthiola incana*, petals (Hemleben, 1981), blue crystals in larkspur, *Consolida ambigua* (Asen *et al.*, 1975) and red crystals in mung bean hypocotyls (Nozzolillo and Ishikura, 1988).

Although the occurrence of these structures in petal cells has received little attention to date, the occurrences of these structures in other parts of plants has received more study. The structures that have been observed to occur in leaves of various Brassicaceae (Small and Pecket, 1982; Nozzolillo *et al.*, 1995), in the tubers of sweet potato, *Ipomea batatus* (Nozue *et al.*, 1997) and in grapes (Cormier and Do, 1993) have received more attention. Early reports had described these as membrane bound 'anthocyanoplasts', membrane bound organelles that provide intense colouration within the vacuoles of mature cells. More recent reports however describe that in a number of species these aggregations of pigment are a combination of anthocyanins and specific proteins that possess neither a membrane boundary nor internal structure. Studies of the aggregations that have been observed in *Ipomoea batatus* cells has revealed that in this case they are a combination of anthocyanin and a specific protein partner termed VP24. The gene encoding this protein has been cloned and it appears that this protein is part of a much larger 50 kDa protein and may well be a N-terminal cleavage product. The localisation and function of the C-terminal protein is unknown at this time (Wenxin *et al.*, 2000).

Investigations into a similar phenomenon that has been observed in the epidermal cells of lisianthus (*Eustoma grandiflorum*) petals have revealed a similar association of protein with anthocyanins (Markham *et al.*, 2000a). These aggregations were named anthocyanic vacuolar inclusions (AVIs) and are vacuolar localised, non-membrane bound, highly pigmented bodies (see Plate 3). In this case however the complex was found to be substantially more stable than VP24 and it has been shown that the association is between three major acidic proteins (54 and 33 kDa in size) and a specific subset of the anthocyanins produced in these cells. Interestingly this association is absolutely specific for the acylated diglycoside anthocyanins and represents a unique copigmentation situation for the bound anthocyanins. More recent work has indicated that in related cultivars of lisianthus the association is not limited to the acylated diglycosides of delphinidin but also those glycosides of pelargonidin and cyanidin (Winefield and Markham, unpublished). Additionally there is strong evidence to show that floral pigmentation (in this case a pink colouration) in these cultivars results directly from the association of pigment with this complex and absence of the complex would result in the lack of colour due to the vacuolar pH conditions within the vacuoles of the epidermal cells (Winefield and Markham, unpublished). Investigations of a wide range of species, indicates that this phenomenon is more widely spread than is indicated from the literature. Indeed it appears that members of most species under certain conditions are capable of producing these aggregations (Kevin Gould, personal communication). Comparison of the structures seen in *Ipomea* and lisianthus, however, indicate that there may not be one common mechanism for their formation. The exact function of these aggregations within the plants is unclear at this time. While numerous possibilities such as anthocyanin 'sinks' or as methods for intracompartmental sequestration can be thought of, we will have to wait for further experimental data before we can assign a function for the formation of these aggregates.

VI. CONCLUSIONS

The steps controlling and contributing to the production of anthocyanins represents only part of the story in determining the processes that contribute to the *in vivo* function of anthocyanins. Subtle changes to the chemistry can have a major impact on the absorbance spectra of the anthocyanins thereby leading to significant changes to colour and function of these pigments. This realisation has lead to resurgence in investigations into the terminal steps of this pathway as it has become evident that for effective molecular manipulation of the anthocyanins, these considerations have to be taken into account. The relationships between anthocyanin chemistry, stabilisation and the surrounding environment into which these compounds are placed, has revealed important findings for the potential use of anthocyanins in a range to industrial applications, such as food colourants. This has also tuned our thinking of how we might manipulate the pathway *in vivo* for the rational and effective production of new anthocyanin types to suit various applications.

While the driving force behind a significant proportion of this research has been the desire to manipulate the pathway to our own ends, this research has also challenged our understanding of what is a very important part of this pathway. As we can see from mutations to the genes involved in these latter stages, inappropriate substitution and localisation of anthocyanins can have deleterious effects to the plant. Proper function of anthocyanins is dependant on both the appropriate chemical substitution of the anthocyanin and the location into which it is finally deposited. Both the function and the sequestration are interwoven processes that reflect the evolutionary path that plants have utilised to both produce and detoxify a wide range of secondary metabolites. The characteristically wide substrate specificities of those enzymes involved in the latter stages of this pathway is a striking example of how the plants have evolved to cope with the plethora of compounds produced and encountered during the lifetime of a plant.

It is evident that while we currently have a very good understanding of the 'players' in the formation and sequestration of anthocyanins, there is still a large amount of work to be done to fully understand how these biosynthetic genes, gene regulators, transporters and potential anthocyanin-binding proteins co-ordinately act to deposit and correctly 'display' anthocyanins in plants so that they can enact their proper functions.

ACKNOWLEDGEMENTS

The author would like to thank Virginia Walbot and the members of her laboratory for sharing information about *Bz2* and stimulating discussions about sequestration in general. The author would also like to thank Kevin Davies and the members of the Plant Pigments Group at Crop and Food Research NZ Ltd for their support.

REFERENCES

Akashi, T., Fukuchi-Mizutani, M., Aoki, T., Ueyama, Y., Yonekura-Sakakibara, K., Tanaka, Y., Kusumi, T. and Ayabe, S-i. (1999). Molecular cloning and biochemical characterization of a novel cytochrome P450, flavone synthase II, that catalyses direct conversion of flavanones to flavones. *Plant Cell Physiology* **40**, 1182–1186.

Alfenito, M. R., Souer, E., Goodman, C. D., Buell, R., Mol, J., Koes, R. and Walbot, V. (1998). Functional complementation of anthocyanin sequestration in the vacuole by widely divergent glutathione S-transferases. *The Plant Cell* **10**, 1135–1149.

Asen, S., Stewart, R. N. and Norris, K. H. (1975). Anthocyanin, flavonol copigments, and pH responsible for larkspur flower colour. *Phytochemistry* **14**, 2677–2682.

Bloor, S. J. (1997). Blue flower colour derived from flavovol-anthocyanin co-pigmentation in *Ceanothus papillosus*. *Phytochemistry* **45**, 1399–1405.

Bloor, S. J. (1999). Novel pigments and copigmentation in blue marguerite daisy. *Phytochemistry* **50**, 1395–1399.

Bloor, S. J. and Falshaw, R. (2000). Covalently linked anthocyanin-flavonol pigments from blue Agapanthus flowers. *Phytochemistry* **53**, 575–579.

Brouillard, R. and Dangles, O. (1994). Flavonoids and flower colour. *In* "The Flavonoids – Advances in Research Since 1986" (J. B. Harborne, ed.), pp. 565–588. Chapman and Hall, London.

Brown, M. H., Paulsen, I. T. and Skurray, R. A. (1999). The multidrug efflux protein NorM is a prototype of a new family of transporters. *Molecular Microbiology* **31**, 393–395.

Brugliera, F., Barri-Rewell, G., Holton, T. A. and Mason, J. G. (1999). Isolation and characterization of a flavonoid 3′-hyroxylase cDNA clone corresponding to the *HT1* locus of *Petunia hybrida*. *The Plant Journal* **19**, 441–451.

Burbulis, I. E. and Winkel-Shirley, B. (1999). Interactions among enzymes of the *Arabidopsis* flavonoid biosynthetic pathway. *The Proceedings of the National Academy of Sciences, USA* **96**, 12929–12934.

Coleman, J. O. D., Mechteld Blake-Klaff, M. A. and Emyr Davies, T. G. (1997). Detoxification of xenobiotics by plants: chemical modification and vacuolar compartmentation. *Trends in Plant Science* **2**, 144–151.

Cormier, F. and Do, C. B. (1993). XXVII *Vitus vinifera* L. (grapevine): *in vitro* production of anthocyanins. *In* "Biotechnology in Agriculture and Forestry" (Y. P. S. Bajaj, ed.), pp. 373–386. Springer Verlag, Berlin.

Debeaujon, I., Peeters, A. J. M., Léon-Kloosterzeil, K. M. and Koorneef, M. (2001). The *TRANSPARENT TESTA 12* gene of arabidopsis encodes a multidrug secondary transporter-like protein required for flavonoid sequestration in vacuoles of the seed coat endothelium. *The Plant Cell* **13**, 853–871.

Edwards, R., Dixon, D. P. and Walbot, V. (2000). Plant glutathione *S*-transferases: enzymes with multiple functions in sickness and in health. *Trends in Plant Science* **5**, 193–198.

Figueiredo, P., Elhabiri, M., Toki, K., Saito, N., Dangles, O. and Brouillard, R. (1996). New aspects of anthocyanin complexation. Intramolecular copigmentation as a means for colour loss? *Phytochemistry* **41**, 301–308.

Fujiwara, H., Tanaka, Y., Fukui, Y., Nakao, M., Ashikari, T. and Kusumi, T. (1997). Anthocyanin 5-aromatic acyltransferase from *Gentiana triflora*. Purification, characterization and its role in anthocyanin biosynthesis. *European Journal of Biochemistry* **249**, 45–51.

Grotewold, E. (2001). Subcellular trafficking of phytochemicals. *Recent Research Developments in Plant Physiology* **2**, 31–48.

Hemleben, V. (1981). Anthocyanin carrying structures in specific genotypes of *Matthiola incana* R. Br. *Z. Naturforsch* **36c**, 925–927.

Holton, T. A. and Cornish, E. (1995). Genetics and biochemistry of anthocyanin biosynthesis. *The Plant Cell* **7**, 1071–1083.

Holton, T. A., Brugliera, F., Lester, D. R., Tanaka, Y., Hyland, C. D., Menting, J. G. T., Lu, C., Farcy, E., Stevenson, T. W. and Cornish, E. C. (1993). Cloning and expression of cytochrome P450 genes controlling flower colour. *Nature* **366**, 276–279.

Jones, P. and Vogt, T (2000). Glycosyltransferases in secondary plant metabolism: tranquilizers and stimulant controllers. *Planta* **213**, 164–174.

Keegstra, K. and Raikhel, N. (2001). Plant glycosyltransferases. *Current Opinion in Plant Biology* **4**, 219–224.

Klien, M., Weissenböck, G., Dufaud, A, Gaillard, C., Kreuz, K. and Martinoia, E. (1996). Different energization mechanisms drive the vacuolar uptake of a flavonoid glucoside and herbicide glucoside. *Journal of Biological Chemistry* **271**, 29666–29671.

Klien, M., Martinoia, E., Hoffman-Thoma, G. and Weissenböck, G. (2000). A membrane-potential dependant ABC-like transporter mediates the vacuolar uptake of rye flavone glucuronides: regulation of glucuronide uptake by glutathione and its conjugates. *The Plant Journal* **21**, 289–304.

Klien, M., Martinoia, E., Hoffman-Thoma, G. and Weissenböck, G. (2001). The ABC-like vacuolar transporter for rye mesophyll flavone glucuronides is not species-specific. *Phytochemistry* **56**, 153–159.

Kubo, H., Nozue, M., Kawasaki, K. and Yasuda, H. (1995). Intravacuolar spherical bodies in *Polygonum cuspidatum*. *Plant and Cell Physiology* **36**, 1453–1458.

Li, Y., Baldauf, S., Lim, E-K. and Bowles D. J. (2001). Phylogenetic analysis of the UDP-glycosyltransferase multigene family of *Arabidopsis thaliana*. *Journal of Biological Chemistry* **276**, 4338–4343.

Lu, Y-P., Li, Z-S. and Rea, P. A. (1997). *AtMRP1* gene of *Arabidopsis* encodes a glutathione S-conjugate pump: Isolation and functional definition of a plant ATP-binding cassette transporter gene. *The Proceedings of the National Academy of Sciences, USA* **94**, 8243–8248.

Lu, Y-P., Li, Z-S., Drozdowicz, Y. M., Hörtensteiner, S., Martinoia, E. and Rea, P. A. (1998). AtMRP2, an arabidopsis ATP binding cassette transporter able to transport glutathione *S*-conjugates and chlorophyll catabolites: functional comparisons with AtMRP1. *The Plant Cell* **10**, 267–282.

Markham, K. R., Gould, K. S., Winefield, C. S., Mitchell, K. A., Bloor, S. J. and Boase, M. R. (2000a). Anthocyanic vacuolar inclusions – their nature and significance in flower colouration. *Phytochemistry* **55**, 327–336.

Markham, K. R., Ryan, K. G., Gould, K. S. and Rickards, G. K. (2000b). Cell wall sited flavonoids in lisianthus flower petals. *Phytochemistry* **54**, 681–687.

Markham, K. R., Gould, K. S. and Ryan, K. G. (2001). Cytoplasmic accumulation of flavonoids in flower petals and its relevance to yellow flower colouration. *Phytochemistry* **58**, 403–413.

Marrs, K. A. (1996). The function and regulation of glutathione S-transferases in plants. *Annual Review of Plant Physiology and Plant Molecular Biology* **47**, 127–158.

Marrs, K. A. and Walbot, V. (1997). Expression and RNA splicing of the maize glutathione *S*-transferase *Bronze2* gene is regulated by cadium and other stresses. *Plant Physiology* **113**, 93–102.

Marrs, K. A., Alfenito, M. R., Lloyd, A. M. and Walbot, V. (1995). A glutathione *S*-transferase involved in vacuolar transfer encoded by the maize gene *Bronze-2*. *Nature* **375**, 397–400.

Martenss, S. and Forkmann, G. (1999). Cloning and expression of flanone synthase II from *Gerbera* hybrids. *The Plant Journal* **20**, 611–618.

McGonigle, B., Keeler, S. J., Lau, S.-M., Koeppe, M. K. and O'Keefe, D. P. (2000). A genomics approach to the comprehensive analysis of the glutathione *S*-transferase gene family in soybean and maize. *Plant Physiology* **124**, 1105–1120.

Mol, J., Grotewold, E. and Koes, R. (1998). How genes paint flowers and seeds. *Trends in Plant Science* **3**, 212–217.

Mol, J., Cornish, E., Mason, J. and Koes, R. (1999). Novel coloured flowers. *Current Opinion in Biotechnology* **10**, 198–201.

Mueller, L. A., Goodman, C. D., Silady, R. A. and Walbot, V. (2000). AN9, a Petunia gluthathione *S*-transferase required for anthocyanin sequestration, is a flavonoid-binding protein. *Plant Physiology* **123**, 1561–1570.

Nozue, M., Yamada, K., Nakamura, T., Kubo, H., Kondo, M. and Nishimura, M. (1997). Expression of a vacuolar protein (VP24) in anthocyanin-producing cells of sweet potato in suspension culture. *Plant Physiology* **115**, 1065–1072.

Nozzolillo, C. and Ishikura, N. (1988). An investigation of the intracellular site of anthocyanoplasts using isolated protoplasts and vacuoles. *Plant Cell Reports* **7**(6), 389–392.

Nozzolillo, C., Andersen, J. and Warwick, S. (1995). Anthocyanoplasts in the Brassicaceae: does their presence serve as a chemotaxonomic marker within the family? *Polyphenols Actualites* **12**, 25–26.

Ralston, E. J., English, J. J. and Dooner, H. K. (1988). Sequence of three bronze alleles of maize and correlation with the genetic fine structure. *Genetics* **119**, 185–197.

Rea, P. A. (1999). MRP subfamily ABC transporters from plants and yeast. *Journal of Experimental Botany* **50**, 895–913.

Rea, P. A., Li, Z-S., Lu, Y-P. and Drozdowicz, Y. M. (1998). From vacuolar GS-X pumps to multispecific ABC transporters. *Annual Review of Plant Physiology and Plant Molecular Biology* **49**, 727–760.

Shirley, B. W. (1996). Flavonoid biosynthesis: 'new' functions for an 'old' pathway. *Trends in Plant Science* **1**, 377–382.

Small, C. J. and Pecket, R. C. (1982). The ultrastructure of anthocyanoplasts in red-cabbage. *Planta* **154**, 97–99.

Vogt, T. and Jones, P. (2000). Glycosyltransferases in plant natural product synthesis: characterization of a supergene family. *Trends in Plant Science* **5**, 380–386.

Walczak, H. A. and Dean, J. V. (2000). Vacuolar transport of the glutathione conjugate of *trans*-cinnamic acid. *Phytochemistry* **53**, 441–446.

Wenxin, X., Moriya, K., Yamada, K., Nishimura, M., Shioiri, H., Kojima, M. and Nozue, M. (2000). Detection and characterization of a 36-kDA peptide in C-terminal region of a 24-kDA vacuolar protein (VP24) precursor in anthocyanin-producing sweet potato cells in suspension culture. *Plant Science* **160**, 121–128.

Winkel-Shirley, B. (2001). Flavonoid biosynthesis. A colorful model for genetics, biochemistry, cell biology, and biotechnology. *Plant Physiology* **126**, 485–493.

Yamazaki, M., Gong, Z., Fukuchi, M., Fukui, Y., Tanaka, Y., Kusumi T. and Saito, K. (1999). Molecular cloning and biochemical characterization of a novel anthocyanin 5-O-glycosyltransferase by mRNA differential display for plant forms regarding anthocyanin. *Journal of Biological Chemistry* **274**, 7405–7411.

Yasuda, H. (1974). Studies on 'bluing effect' in the petals of red rose II. Observation on the development of the tannin body in the upper epidermal cells of bluing petals. *Cytologica* **39**, 107–112.

Molecular Genetics and Control of Anthocyanin Expression

BRENDA WINKEL-SHIRLEY

Department of Biology, Virginia Tech, Blacksburg, VA 24061-0406, USA

ABSTRACT

The flavonoid pathway leading to anthocyanins has been well-characterized in a number of model systems, including maize, petunia, snapdragon and, more recently, *Arabidopsis*. Genetic approaches have identified many of the regulatory and structural genes required for the synthesis of these important plant pigments. Analysis of mutant lines has also provided evidence for biological functions of flavonoids other than those related to pigmentation. In all species examined to date, expression of the pathway is highly regulated in response to developmental and environmental cues and occurs, at least in part, at the level of transcription. The assembly of flavonoid enzymes into a membrane-bound macromolecular complex may provide yet another important mechanism for controlling pathway flux and determining the relative levels of specific endproducts. Understanding these diverse aspects of flavonoid metabolism is crucial for efforts to engineer this system for horticultural, nutritional, and agronomic improvement of plants.

Advances in Botanical Research Vol. 37
incorporating Advances in Plant Pathology
ISBN 0-12-005937-1

I. INTRODUCTION

One of the most conspicuous characteristics of flavonoids is their contribution to the beautiful and diverse pigmentation that characterizes the plant world (reviewed in Mol *et al.*, 1998; Forkmann and Martens, 2001; Winkel-Shirley, 2001a). Three types of flavonoids are synthesized by virtually all higher plants: anthocyanins, which are responsible for many of the red, blue and purple colors in flowers, fruits, vegetables, and leaves; flavonols, which provide yellow color and also function as copigments to modify anthocyanin coloration; and proanthocyanidins or condensed tannins, polymeric molecules that are responsible for the brown pigmentation of many plant seeds. In addition, specialized forms of flavonoids are produced in a number of species, such as aurones, which also provide yellow coloration, and the 3-deoxyanthocyanins, that provide red pigmentation in a small number of plants, including maize and sorghum. Two other important classes of flavonoids, the flavanones and isoflavonoids, are also synthesized by a few plant species and do not generally contribute to plant pigmentation but, like many of the flavonoids, play a number of other essential roles in plants. In fact, it has been suggested that the earliest function of flavonoids was in UV protection and that the pathway subsequently evolved to provide compounds with diverse functions in pigmentation, signaling, and stress protection (Stafford, 1991). The biochemical steps leading to the various flavonoid compounds are outlined in Fig. 1.

The striking pigmentation provided by flavonoids has formed the basis for extensive classical genetic studies in a variety of plant species, starting with the experiments of Gregor Mendel and the subsequent work of Barbara McClintock. More recently, genetic approaches have identified loci defining both structural and regulatory genes required for flavonoid biosynthesis. In horticultural species such as petunia, snapdragon, carnation, matthiola, and morning glory, flavonoid mutants have been identified largely on the basis of altered flower color (Geissman *et al.*, 1956; Forkman, 1977; Leweke and Forkmann, 1982; Holton and Cornish, 1955; Mol *et al.*, 1998; Clegg and Durbin, 2000). In plants like maize, barley, and *Arabidopsis*, seed coat coloration has provided a phenotype for the identification of flavonoid mutants (von Wettstein *et al.*, 1980; Holton and Cornish, 1995; Shirley *et al.*, 1995; Mol *et al.*, 1998). Table I gives a summary of loci for flavonoid enzymes in maize, barley, *Arabidopsis*, petunia, and snapdragon, the five systems in which genetic approaches to the study of this pathway have been the most extensive.

The isolation of genes identified by flavonoid regulatory loci has relied heavily on transposon and T-DNA tagging approaches and a significant number of these genes have now been cloned and characterized, particularly in maize, petunia, and *Arabidopsis* (reviewed in

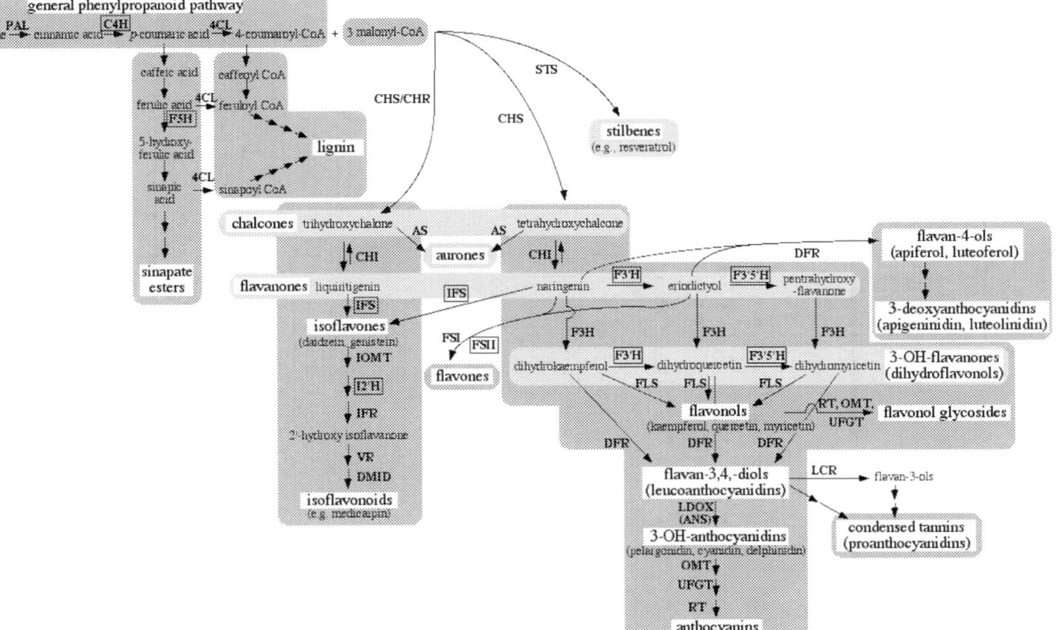

Fig. 1. Schematic of the branch pathways of general phenylpropanoid metabolism leading to flavonoids, isoflavonoids, sinapate esters and lignin. The names of the major classes of compounds are in white boxes; cytochrome P450 enzymes are indicated with open boxes. Enzyme names are abbreviated as follows: anthocyanidin synthase (ANS), aureusidin synthase (AS), cinnamate-4-hydroxylase (C4H), chalcone isomerase (CHI), chalcone reductase (CHR), chalcone synthase (CHS), 4-coumaroyl:CoA-ligase (4CL), dihydroflavonol 4-reductase (DFR), 7,2′-dihydroxy, 4′-methoxyisoflavanol dehydratase (DMID), flavanone 3-hydroxylase (F3H), flavone synthase (FSI and FSII), flavonoid 3′ or 3′5′ hydroxylase (F3′H, F3′5′H), ferulate 5-hydroxylase (F5H), isoflavone *O*-methyltransferase (IOMT), isoflavone reductase (IFR), isoflavone 2′-hydroxylase (I2′H), isoflavone synthase (IFS), leucoanthocyanidin dioxygenase (LDOX), leucoanthocyanidin reductase (LCR), *O*-methyltransferase (OMT), phenylalanine ammonia-lyase (PAL), rhamnosyl transferase (RT), stilbene synthase (STS), UDP flavonoid glucosyl transferase (UFGT), vestitone reductase (VR).

TABLE I

Genetic loci for flavonoid biosynthetic genes

Enzyme	Maize[a]	Barley[b]	Arabidopsis[c]	Petunia[a,f]	Snapdragon[a]
chalcone synthase	*c2*, *whp*	–	*tt4*	*wha*	*nivea*
chalcone isomerase	–	–	*tt5*	*po*	–
flavanone 3-hydroxylase	–	–	*tt6*	*an3*	*incolorata*
flavonoid 3′-hydroxylase	*pr*	–	*tt7*	*ht1*	*eosina*
flavonoid 3′5′-hydroxylase	–	–	–	*hf1*, *hf2*	–
flavonol synthase	–	–	*fls1::En*	–	–
dihydroflavonol reductase	*a1*	*ant18*	*tt3*	*an6*	*pallida*
leucoanthocyanidin reductase	–	*ant19*	*ban/ast*	–	–
leucoanthocyanidin dioxygenase/anthocyanidin synthase	*a2*	–	*tt18*[d]	–	*candica*
anthocyanin methyltransferase	–	–	–	*mf1/mf2* *mt1/mt2*	–
flavonoid 3-glucosyl transferase	*bz1*	–	*fgt–1*[e]	–	–
anthocyanin rhamnosyltransferase	–	–	–	*rt*	–
anthocyanin acyltransferase	–	–	–	*gf*	–

[a]Holton and Cornish (1995)
[b]Jende-Strid (1991)
[c](Winkel-Shirley, 2001a),
[d]Atsushi Taraka, personal communication.
[e]Baxter-Burrell *et al.* (2000)
[f]Kroon *et al.* (1994); de Vetten *et al.* (1999); Napoli *et al.* (1999).

Winkel-Shirley, 2001b). The isolation of structural genes, on the other hand, has been accomplished largely using biochemical approaches, thanks to the high-level, and in some cases inducible, expression of flavonoid enzymes in tractable systems such as parsley and bean cell suspension cultures and the flowers of matthiola, petunia, and carnation (reviewed in Winkel-Shirley, 2001b). Even where mutants were not needed for gene isolation, these lines have provided useful confirmation of gene function and have also led to the identification of biological roles for flavonoids other than in pigmentation. Genes for the isoflavonoid branch pathway, which produces unpigmented flavonoids and for which few, if any, mutants are available, have also been isolated based largely on biochemical and, more recently, genomic approaches in pea, soybean and alfalfa (reviewed in Winkel-Shirley, 2001b).

The long-range goal of much of this work has been not only to understand the regulation and enzymology of flavonoid metabolism, but also to apply this knowledge for practical purposes. There has been a longstanding interest in engineering flower color using the flavonoid pathway, which has provided a strong impetus for deciphering the complexities of enzymes and other factors that contribute to this remarkable phenomenon (reviewed in Mol *et al.*, 1998, 1999; Tanaka *et al.*, 1998; Forkmann and Martens, 2001). The nutritional aspects of flavonoids have been a more recent focus, with the goal of altering the amounts or types of flavonoids produced in crops for human consumption and animal forage (Morris and Robbins, 1997; Dixon and Steele, 1999). For example, Muir *et al.* (2001) have used a petunia CHI gene to elevate the levels of flavonol antioxidants in tomato fruit. In addition, efforts have been directed at engineering all or part of the pathway leading to isoflavonoids, which are also believed to have important health-promoting activities, in plant species that do not normally synthesize these compounds. The success of initial experiments, involving the expression of isoflavone synthase in *Arabidopsis* and tobacco and, together with chalcone reductase and a C1/R chimeric transcription factor, in maize indicates the feasibility of this approach (Jung *et al.*, 2000; Yu *et al.*, 2000). Engineering condensed tannins has also been of interest with regard to forage crops, as these compounds can be beneficial in reducing pasture bloat but also have antinutrient properties (Morris and Robbins, 1997; Gruber *et al.*, 1999). Some progress has been made toward this goal by using flavonoid enzymes from different species to alter the composition or levels of condensed tannins in transgenic plants. Metabolic engineering is not limited to the use of enzymes, of course. As more becomes known about the complex transcriptional regulation of flavonoid metabolism, there is also the possibility of using transcription factors to coordinately upregulate entire branch pathways, as suggested by Grotewold and colleagues (Grotewold *et al.*, 1998; Braun *et al.*, 2000).

II. USE OF MUTANTS TO IDENTIFY BIOLOGICAL FUNCTIONS OF FLAVONOIDS OTHER THAN IN PIGMENTATION

Genetic approaches have provided insights, not only into the enzymology and regulation of the flavonoid pathway, but also into the diverse functions of its products. For example, the possibility that flavonoids play a role in protecting plants from UV radiation has drawn a great deal of interest over the years, even more so with recent evidence for stratospheric ozone depletion. Flavonoids generally absorb light in the 280–315 nm region, accumulate to high levels in the vacuoles of epidermal cells, and are synthesized at increased levels in response to UV light, consistent with a role for flavonoids in protecting underlying photosynthetic tissues from UV damage (Kootstra, 1994; Landry *et al.*, 1995; Harborne and Williams, 2000). Although these features have been known for some time, the first direct evidence that flavonoids contribute to UV protection came from experiments in *Arabidopsis* showing that mutants defective in flavonoid biosynthesis were hypersensitive to UV-B irradiation (Li *et al.*, 1993). Interestingly, the chalcone isomerase (CHI) mutant, *tt5*, was found to be more sensitive than the chalcone synthase (CHS) mutant, *tt4*, despite the fact that it was blocked one step later in the biosynthetic pathway (Fig. 1). This appeared to be due to the fact that a second class of UV-absorbing compounds, the sinapate esters, were elevated in *tt4*, but reduced in *tt5* relative to wild-type. In fact, sinapate levels increase in wild-type and *tt4* plants exposed to UV radiation, but do not increase in the *tt5* mutant (Ormrod *et al.*, 1995), suggesting some type of communication between these two pathways at the level of naringenin biosynthesis. Analysis of the *Arabidopsis fah1* mutant, which is defective in sinapate ester biosynthesis and even more sensitive than *tt5* to UV damage based on measures such as foliar injury and lipid and protein oxidation, led to the suggestion that sinapate esters may be more important than flavonoids in UV protection (Landry *et al.*, 1995). This possibility has been borne out in recent studies of the effects of UV-B radiation on turnover of components of the Photosystem II reaction center (Booij-James *et al.*, 2000). Nonetheless, it is clear that flavonoids do contribute to some degree to UV protection in plants based on a number of other studies linking flavonoids with the inhibition of UV-induced DNA damage, both in *in vitro* experiments and in comparative analyses of maize plants (Kootstra, 1994; Stapleton and Walbot, 1994). Related to this, Feild *et al.* (2001) have recently suggested that anthocyanins in senescing leaves serve to reduce the risk of photo-oxidative damage to leaf cells, ensuring the efficient retrieval of nutrients.

Genetic approaches have also provided insights into some unexpected functions of flavonoids, such as the role of these compounds in male fertility. This phenomenon was first identified in the maize mutant, *whp*, which results in white pollen when present in a *c2* background (Coe *et al.*, 1981). These two loci specify chalcone synthase (CHS) activity, which catalyzes the first committed step in flavonoid biosynthesis. Therefore, a *whp;c2* line produces no flavonoids of any kind and the pollen is white rather than the normal yellow color. The finding that these plants were not self-fertile and that crosses made using white pollen were unsuccessful, pointed to a role for flavonoids in maintaining male fertility. A similar phenotype was subsequently described in transgenic petunia in association with cosuppression or antisense inhibition of CHS activity (Taylor and Jorgensen, 1992; van der Meer *et al.*, 1992) and later in an induced CHS mutant (Napoli *et al.*, 1999). However, petunia *po* mutants, which lack CHI expression, are fully fertile; this could be because the *po* mutant is leaky or because the reaction catalyzed by CHI can occur spontaneously (van Tunen *et al.*, 1991). Biochemical complementation experiments in petunia demonstrated that it was the flavonol aglycones that could restore normal germination and pollen tube growth (Mo *et al.*, 1992; Ylstra *et al.*, 1992). Because no flavonol aglycones could be found in either stigmas or anthers at any stage of development, it was suggested that a flavonol glycosidase may function to convert inactive glycosides into the active aglycone form (Pollak *et al.*, 1993). Further analysis also indicated that flavonols might play a structural role in the membranes of the male gametophyte, explaining the observation that the tip of pollen tubes from flavonol-deficient lines burst during the elongation process (Ylstra *et al.*, 1994). One other puzzling finding was that this phenomenon was not universal; *Arabidopsis* mutants defective for CHS activity remain fully fertile (Burbulis *et al.*, 1996; Ylstra *et al.*, 1996). To this day, the precise mechanism by which flavonol aglycones are generated, how these compounds contribute to pollen germ tube integrity in certain plant species, and whether other compounds play this role in plants such as *Arabidopsis* all remain to be elucidated.

The role of flavonoids in interactions between plants and other organisms is certainly well established. There is an extensive literature on the role of flavonoids in plant–animal interactions, including evidence that in specific plants these compounds act as feeding deterrents for insects and animals (reviewed in Harborne and Williams, 2000). There are also numerous examples in which the flavonoid composition of particular plants are used as identifiers by either detrimental or beneficial organisms, including the use of flavonol glycosides in leaves to identify plant foods for oviposition of moths and butterflies (Harborne and Williams, 2000) or the use of flower and fruit color cues for pollinators and seed dispersers (Willson and Whelen, 1990; Weiss, 1991). This also extends to interactions of microbes and plants, in which the host range of organisms such as nitrogen-fixing rhizobia is specified by isoflavonoid and flavonoid signals (Fisher and Long, 1992; McKhann *et al.*, 1998) and flavonoids have even been used to stimulate colonization of a non-legume, *Arabidopsis*, by *Azorhizobium caulinodans* (Gough *et al.*, 1997). Plant–plant interactions can also involve flavonoids as inducers or recognition signals, as in the case of the plant parasites, *Triphysaria versicolor* (Albrecht *et al.*, 1999) and *Cuscuta subinclusa* (Kelly, 1990) and in contact coiling of pea tendrils (Jaffe and Galston, 1967). On the other hand, flavonoids do not appear to be required for host–parasite interaction in the case of *Orobanche* (Westwood, 2000). In addition, while interactions with arbuscular mycorrhizae have been shown to induce flavonoid biosynthetic genes in the host plant and flavonol aglycones can stimulate mycorrhizal hyphal growth, flavonoids do not appear to be required for the establishment of mycorrhizal interactions (Harrison and Dixon, 1994; Bécard *et al.*, 1995; Blee and Anderson, 1996; Buee *et al.*, 2000). The ways in which flavonoids contribute to the interactions of plants with other organisms are clearly complex and much remains to be learned in this fascinating area of plant biology.

Jacobs and Rubery (1988) were the first to suggest that flavonoids function to regulate polar auxin transport by inhibiting the auxin efflux carrier. This concept remained controversial and unsubstantiated for many years, until recent work with *Arabidopsis* mutants defective in flavonoid biosynthesis began to provide new support for the idea. In one study, Murphy *et al.* (2000) showed that seedlings of the *Arabidopsis* CHS mutant, *tt4*, have an altered pattern of auxin distribution, and that significantly more radiolabeled auxin is expelled into the medium from roots of *tt4* seedlings than from wild-type seedlings. Brown *et al.* (2001) have also shown that auxin transport is elevated in *Arabidopsis* mutants defective in flavonoid biosynthesis and that these plants have numerous phenotypic characters consistent with altered auxin transport. In this regard it is intriguing that CHS and CHI have been localized to the apical end of cortex cells in the root elongation zone (Saslowsky and Winkel-Shirley, 2001), opposite from where AtPIN2, a component of the efflux carrier, occurs (Müller *et al.*, 1998). This suggests the possibility that the strikingly asymmetric localization of the flavonoid biosynthetic enzymes might be important for controlling the subcellular distribution of flavonoids, perhaps related to a role for these compounds in controlling polar auxin transport.

III. TRANSCRIPTIONAL REGULATION OF FLAVONOID BIOSYNTHESIS

Flavonoid biosynthesis is regulated in response to an enormous diversity of intrinsic and environmental signals (reviewed in Chalker-Scott, 1999).

These expression patterns are consistent with many of the known and suspected roles of flavonoids in plants, but also point to additional, as-yet-uncharacterized functions for these compounds. There is extremely well-defined developmental, tissue- and even cell-specific control over the accumulation of flavonoids in all plants. Some of this is clearly linked to functions such as recruitment of pollinators to flowers or of nitrogen-fixing rhizobia to roots, although the function of the transient accumulation of anthocyanins that is observed in many plants during seedling development remains unknown (Chalker-Scott, 1999), and is partly the subject of this volume. Flavonoid synthesis can also be induced in response to a wide range of stresses including cold, heat, salt, drought, wounding, and nitrogen and phosphorus deficiency, perhaps indicating that flavonoid synthesis is part of a general stress response in plants. The specific role of this induction is also poorly understood, although it is believed that flavonoids may function, at least in part, as antioxidants by scavenging free radicals (Harborne and Williams, 2000; Rice-Evans, 2001). In most, if not all, plant systems, it is clear that transcriptional control is an important element governing expression of the flavonoid pathway during development or in response to external cues. How these signals are integrated to control gene expression is a topic of ongoing investigation.

The flavonoid pathway provided one of the earliest examples of coordinate enzyme expression in plants, as described by Dooner *et al.* (1983) based on studies in maize. Biochemical evidence for simultaneous expression of flavonoid enzymes was quickly corroborated with genetic evidence for regulatory loci that simultaneously controlled the expression of multiple flavonoid genes. This model has been extended to a number of other plant species, and a significant number of regulatory genes have now been isolated and characterized in maize, petunia, snapdragon and *Arabidopsis*. The original model of coordinate control of gene expression has been refined to incorporate the complexities that actually characterize flavonoid gene regulation. In many plants there appears to be a distinct regulatory mechanism for flavonoid genes required for early biosynthetic steps leading to flavonols and the precursors for proanthocyanidins and anthocyanins, and for those catalyzing later steps. This is illustrated by the regulatory mutants *an1* and *an11* in petunia, which affect expression of ‘late’ genes such as dihydroflavonol reductase (*DFR*), anthocyanin rhamnosyl transferase, and anthocyanin methyltransferases, but not the ‘early’ genes, *CHS*, *CHI*, and flavanone 3-hydroxylase (*F3H*) (Quattrocchio *et al.*, 1993) (Fig. 1). Similarly, the *TTG1* locus in *Arabidopsis* is known to be required for expression of the ‘late’ genes, *DFR*, leucoanthocyanidin dioxygenase (*LDOX*), and banyuls, but not for the ‘early’ genes, *CHS*, *CHI*, *F3H*, flavonol synthase, or flavonoid 3′-hydroxylase (*F3′H*) (Pelletier and Shirley, 1995; Shirley and Hwang, 1995; Pelletier *et al.*, 1997; Nesi *et al.*, 2000). However, in snapdragon flowers, analysis of the regulatory mutants, *eluta* and *delila*, uncovered a different division, with *F3H* belonging to the 'late' group of genes in this species (Martin *et al.*, 1991; Jackson *et al.*, 1992). Analysis of additional flower-specific *myb* genes in a yeast expression system showed activation of *PAL*, *CHI* and *F3H* promoters, but not the *CHS*, *DFR* and *LDOX* promoters (reviewed in Mol *et al.*, 1998), suggesting that either the situation is more complex in snapdragon, or that it has not yet been fully characterized in other plants. Maize appears to use a somewhat different mechanism, in which specific transcription factors coordinately control expression of subsets of genes required for the synthesis of distinct endproducts. More specifically, the product of the *P* gene specifically activates the *CHS*, *CHI*, and *DFR* genes in maize floral organs, leading to the synthesis of 3-deoxyanthocyanins (reviewed in Mol *et al.*, 1998), while R and C1 together activate all of the flavonoid biosynthetic genes, with the exception of CHI, leading to the synthesis of flavonols and anthocyanins in a variety of tissues (Braun *et al.*, 2000). Screens for novel seed color mutants in maize indicate that additional, as-yet-uncharacterized regulatory genes are involved, such as *PAC1*, which is required for C1 and P activation of flavonoid gene expression (Selinger and Chandler, 1999). The situation in barley may be similar to that in maize, as *ant 13* mutants result in a substantial reduction of both *CHS* and *DFR* mRNA (Jende-Strid, 1991). There is also evidence that mutations that impact flavonoid accumulation may do so through indirect means involving effects on cell morphogenesis, for example in the case of *Arabidopsis TTG1* locus, which affects not only flavonoid synthesis, but also the development of trichomes and root hairs, and various characteristics of the seed coat (Debeaujon *et al.*, 2000; Western *et al.*, 2001). A similar pleiotrophic phenotype has been identified in *Matthiola incana*, another crucifer, but has not yet been reported in other plant species (Kappert, 1949). This is yet anther complex area of flavonoid biology in which much remains to be learned.

IV. REGULATION OF FLUX BY PATHWAY INTERMEDIATES

In addition to regulation of the pathway in response to specific internal and external cues, flavonoid biosynthesis also appears to be extremely sensitive to the rate of flux through the system and the accumulation of phenylpropanoid intermediates. The best-studied examples involve regulation by *trans*-cinnamic acid, the product of the phenylalanine ammonia-lyase (PAL) reaction. This compound has been shown to inhibit accumulation of transcripts from PAL genes in French bean (*Phaseolus vulgaris*) cell suspension cultures, suggesting that this intermediate, or a derivative, is a feedback regulator of *PAL* gene transcription (Mavandad *et al.*, 1990). Although the levels of cinnamic acid used in this study were much higher than physiological (Hrazdina and Jensen, 1992), this finding is consistent with inhibitor studies showing that PAL expression is induced by inhibition of PAL activity and repressed by inhibition of C4H (summarized in Blount *et al.*, 2000). The best evidence for feedback inhibition of PAL activity by *trans*-cinnamic acid has come from recent analysis of transgenic plants in which the second enzyme of phenylpropanoid metabolism, cinnamic acid 4-hydroxylase (C4H), was downregulated using sense and antisense expression, again leading to a reduction in PAL activity and the levels of the major products of the sinapate and flavonoid pathways (Blount *et al.*, 2000). This regulation appears to occur, at least in part, at the level of *PAL* gene expression. Interestingly, conversion of *trans*-cinnamic acid to the non-inhibiting *cis* isomer has been suggested as a mechanism underlying the rapid activation of PAL enzyme activity upon UV-B exposure in rye (*Secale cereale*) (Braun and Tevini, 1993). *trans*-Cinnamic acid has also been found to inhibit *CHS* promoter activity when present at high levels in elicited alfalfa cultures, while *trans*-*p*-coumaric acid, the product of C4H, stimulates *CHS* promoter activity (Loake *et al.*, 1991). This stimulation by *p*-coumaric acid appears to operate through the same promoter elements required for developmental and stress induction of *CHS* transcription, indicating that diverse signals control expression of the *CHS* gene via shared mechanisms (Loake *et al.*, 1992). The emerging picture is that key enzymatic steps in phenylpropanoid metabolism gauge flux through the general phenylpropanoid pathway and thereby control the activity of the system as a whole as well as the relative activities of branch pathways.

It is quite likely that feedback regulation of phenylpropanoid metabolism is not limited to intermediates of the general phenylpropanoid pathway, however. It was already shown many years ago that PAL enzyme activity in pea (*Pisum sativum*) is inhibited by the flavonols, quercetin, kaempferol, and quercetin-3-*p*-coumaroyltriglucoside (Attridge *et al.*, 1971). It has also been suggested that flavonols may affect the accumulation of specific flavonoid enzymes, based on studies of *Arabidopsis transparent testa* mutants (Barthet and Winkel-Shirley, unpublished observations; Pelletier *et al.*, 1999). Moreover, mutations that disrupt CHS or the second enzyme of flavonoid biosynthesis, CHI, have been found to affect the levels of sinapate esters in *Arabidopsis*, providing evidence for crosstalk between the flavonoid and sinapate pathways, as discussed earlier (Li *et al.*, 1993). The possibility that downstream products function together with intermediates of phenylpropanoid metabolism to modulate flux through the system clearly merits further investigation.

V. ORGANIZATION OF FLAVONOID METABOLISM AS A MULTI-ENZYME COMPLEX

There are a number of mechanisms by which cellular metabolism can be organized so as to facilitate the coordination of biosynthetic activities. Among these is the organization of enzymes into macromolecular complexes, or metabolons, that provide for high local substrate concentrations, the ability to sequester highly reactive or toxic intermediates from the rest of the cell, and mechanisms for rapid regulation of metabolic activity, among other characteristics (reviewed in Ovádi and Srere, 2000). The organization of phenylpropanoid metabolism as a multi-enzyme complex was first proposed by Helen Stafford in order to address the fundamental problem of competition for common substrates among the multiple branch pathways in this system (Stafford, 1974). The first evidence that this type of organization did indeed exist came from several studies showing close association of PAL and C4H based on channeling of radiolabeled intermediates between active sites (Amrhein and Zenk, 1971; Czichi and Kindl, 1977; Hrazdina and Wagner, 1985b) as well as evidence for distinct metabolic compartmentation or channeling in flavonoid biosynthesis based on unequal utilization of common precursors (Jacques *et al.*, 1977; Margna and Vainjarv, 1981; Stafford, 1981). More recent feeding studies in cell cultures treated with elicitor or expressing a PAL transgene led to the suggestion that there is close association of specific forms of PAL with C4H (Rasmussen and Dixon, 1999). Precursor feeding experiments together with differences in the *in vitro* and *in vivo* regiospecificities of an isoflavone methyltransferase have now also provided evidence for metabolic channeling in the isoflavonoid pathway (Dixon *et al.*, 1998; He and Dixon, 2000).

Cell fractionation and immunolocalization experiments have provided further evidence that the flavonoid pathway is organized as a membrane-associated enzyme complex. Studies in several plant species showed the association of various phenylpropanoid and flavonoid enzymes, most of which are operationally 'soluble' proteins, with microsomal membranes or the endoplasmic reticulum (Fritsch and Grisebach, 1975; Wagner and Hrazdina, 1984; Hrazdina *et al.*, 1987). This led to a model in which cytochrome P450 hydroxylases, such as F3′H, function as membrane anchors that organize the soluble enzymes into multicatalytic complexes (Hrazdina and Wagner, 1985a). Additional support for this model has come from recent work showing co-localization of CHS and CHI in *Arabidopsis* root cells, often at the endoplasmic reticulum, but also in electron-opaque regions (Saslowsky and Winkel-Shirley, 2001). These regions are not observed in an *Arabidopsis* F3′H mutant, consistent with a role for this enzyme in organizing the localization of other flavonoid enzymes. There is also evidence for direct protein interactions among flavonoid enzymes based on two-hybrid, immunoprecipitation and affinity chromatography experiments (Burbulis and Winkel-Shirley, 1999), making a strong case for the organization of the flavonoid pathway as a multi-enzyme complex. What is needed in order to use this information for metabolic engineering of this and other systems is insights into the three-dimensional structure of the complex and identification of the interfaces that direct assembly of the complex. Progress in this direction has been greatly enhanced by the recent solving of crystal structures for CHS and CHI from alfalfa (*Medicago sativa*) (Ferrer *et al.*, 1999; Jez *et al.*, 2000). This has made it possible to use homology modeling to predict the structures of related enzymes from other plant species, providing additional insights into determinants of substrate specificity and the effects of mutant alleles (Dana and Winkel-Shirley, unpublished; Saslowsky *et al.*, 2000; Zheng *et al.*, 2001). The application of additional technologies such as atomic force microscopy and surface plasmon resonance to characterize flavonoid enzyme interactions (Dana, Saslowsky, and Winkel-Shirley, unpublished), together with the new structural information, should lead to progress in defining the molecular basis underlying assembly of the flavonoid enzyme complex. It should ultimately be possible to determine whether phenylpropanoid metabolism consists of a single complex with multiple branch points, or many complexes each dedicated to the synthesis of specific types of end-products or even individual compounds. This information should be extremely useful in efforts to redirect flux or even introduce new activities into this complex metabolic system.

VI. SUMMARY

The flavonoid biosynthetic pathway is one of the best-studied metabolic systems in plants, with a well-characterized biochemical pathway and growing information about the complex regulation of its expression and its subcellular organization. Progress in these areas has been greatly facilitated in recent years by molecular genetic methodologies, which have enabled new approaches to long-standing questions about the structure and regulation of flavonoid genes, the structure of the enzymes, and their organization within the cell. The flavonoid pathway also continues to be a useful system for studying other fundamental biological phenomena, such as cosuppression and paramutation. The availability of mutant lines has provided important insights into new and unexpected biological functions of flavonoid endproducts over the years, and this remains an active area of investigation in many laboratories, as well. There is also a growing awareness of the nutritional and health-promoting qualities of flavonoids in the foods that we consume, which is likely to drive expanded interest in this system. Still, there are many unanswered questions surrounding flavonoids, from a lack of understanding of many of their physiological functions to the mechanism of transport of flavonoid endproducts to their final destinations within the cell, as is discussed in other chapters in this volume. However, the significant progress that has already been made in this area coupled with early successes in engineering flavonoid synthesis for horticultural, nutritional, and agronomic improvement of plants, indicate that the flavonoid pathway will continue to occupy a central place in the study of plant metabolism.

REFERENCES

Albrecht, H., Yoder, J. I. and Phillips, D. A. (1999). Flavonoids promote haustoria formation in the root parasite *Triphysaria versicolor. Plant Physiology* **119**, 585–591.

Amrhein, N. and Zenk, M. H. (1971). Untersuchungen zur Rolle der Phenylalanine Ammonia Lyase (PAL) bei der Regulation der Flavonoidsynthese in Buchweizen (*Fagopyrum Esculentum* Moench). *Zeitschrift Pflanzenphysiologie* **64**, 145–168.

Attridge, T. H., Stewart, G. R. and Smith, H. (1971). End-product inhibition of *Pisum* phenylalanine ammonia-lyase by the *Pisum* flavonoids. *FEBS Letters* **17**, 84–86.

Baxter-Burell, A., Chang, R., Springer, P. and Bailey-Serres, J. (2002). Gene and enchancer trap transposable elements reveal oxygen deprivation-regulated genes and their complex patterns of expression in *Arabidopsis. Annals of Botany* **27**, in press.

Bécard, G., Taylor, L. P., Douds Jr., D. D., Pfeffer, P. E. and Doner, L. W. (1995). Flavonoids are not necessary plant signal compounds in arbuscular mycorrhizal symbioses. *Molecular Plant–Microbe Interactions* **8**, 252–258.

Blee, K. A. and Anderson, A. J. (1996). Defense-related transcript accumulation in *Phaseolus vulgaris* L. colonized by the arbuscular mycorrhizal fungus *Glomus intraradices* Schenck & Smith. *Plant Physiology* **110**, 675–688.

Blount, J. W., Korth, K. L., Masoud, S. A., Rasmussen, S., Lamb, C. and Dixon, R. A. (2000). Altering expression of cinnamic acid 4-hydroxylase in transgenic plants provides evidence for a feedback loop at the entry point into the phenylpropanoid pathway. *Plant Physiology* **122**, 107–116.

Booij-James, I. S., Dube, S. K., Jansen, M. A., Edelman, M. and Mattoo, A. K. (2000). Ultraviolet-B radiation impacts light-mediated turnover of the photosystem II reaction center heterodimer in Arabidopsis mutants altered in phenolic metabolism. *Plant Physiology* **124**, 1275–1284.

Braun, J. and Tevini, M. (1993). Regulation of UV-protective pigment synthesis in the epidermal layer of rye seedlings *(Secale cereale* L. cv. Kustro). *Phytochemistry and Phytobiology* **57**, 318–323.

Braun, E. L., Dias, A. P., Matulnik, T. J. and Grotewold, E. (2000). Transcription factors and metabolic engineering: novel applications for ancient tools. *In* "Recent Advances in Phytochemistry vol. 35: Regulation of phytochemicals by molecular techniques" (J. T. Romeo, J. A. Saunders and B. Matthews, eds), pp. 79–109. Elsevier, New York.

Brown, D. E., Rashotte, A. M., Murphy, A. S., Normanly, J., Tague, B. W., Peer, W. A., Taiz, L. and Muday, G. K. (2001). Flavonoids act as negative regulators of auxin transport *in vivo* in *Arabidopsis thaliana. Plant Physiology* **126**, 524–535.

Buee, M., Rossignol, M., Jauneau, A., Ranjeva, R. and Becard, G. (2000). The presymbiotic growth of arbuscular mycorrhizal fungi is induced by a branching factor partially purified from plant root exudates. *Molecular Plant–Microbe Interactions* **13**, 693–698.

Burbulis, I. E. and Winkel-Shirley, B. (1999). Interactions among enzymes of the *Arabidopsis* flavonoid biosynthetic pathway. *Proceedings of the National Academy of Sciences, USA* **96**, 12929–12934.

Burbulis, I. E., Iacobucci, M. and Shirley, B. W. (1996). A null mutation in the first enzyme of flavonoid biosynthesis does not affect male fertility in Arabidopsis. *The Plant Cell* **8**, 1013–1025.

Chalker-Scott, L. (1999). Environmental significance of anthocyanins in plant stress responses. *Photochemistry and Photobiology* **70**, 1–9.

Clegg, M. T. and Durbin, M. L. (2000). Flower color variation: A model for the experimental study of evolution. *Proceedings of the National Academy of Sciences, USA* **97**, 7016–7023.

Coe, E. H., McCormick, S. M. and Modena, S. A. (1981). White pollen in maize. *Journal of Heredity* **72**, 318–320.

Czichi, U. and Kindl, H. (1977). Phenylalanine ammonia-lyase and cinnamic acid hydroxylase as assembled consecutive enzymes on microsomal membranes of cucumber cotyledons: cooperation and subcellular distribution. *Planta* **134**, 133–143.

Debeaujon, I., Léon-Kloosterziel, K. M. and Koornneef, M. (2000). Influence of the testa on seed dormancy, germination, and longevity in Arabidopsis. *Plant Physiology* **122**, 403–413.

de Vetten, N., ter Horst, J., van Schaik, H.-P., de Boer, A., Mol, J. and Koes, R. (1999). A cytochrome b_5 is required for full activity of flavonoid 3′,5′-hydroxylase, a cylochrome P450 involved in the formation of blue flowers. *Proceedings of the National Academy of Science, USA* **96**, 778–783.

Dixon, R. A. and Steele, C. L. (1999). Flavonoids and isoflavonoids – a gold mine for metabolic engineering. *Trends in Plant Science* **4**, 394–400.

Dixon, R. A. Howles, P. A., Lamb, C., He, X.-Z. and Reddy, J. T. (1998). Prospects for the metabolic engineering of bioactive flavonoids and related phenylpropanoid compounds. *In* "Flavonoids in the Living System" (J. A. Manthey and B. S. Buslig, eds), pp. 55–66. Plenum Press, New York.

Dooner, H. (1983). Coordinate genetic regulation of flavonoid biosynthetic enzymes in maize. *Molecular and General Genetics* **189**, 136–141.

Feild, T. S., Lee, D. W. and Holbrook, N. M. (2001). Why leaves turn red in autumn. The role of anthocyanins in senescing leaves of red-osier dogwood. *Plant Physiology* **127**, 566–574.

Ferrer, J.-L., Jez, J. M., Bowman, M. E., Dixon, R. A. and Noel, J. P. (1999). Structure of chalcone synthase and the molecular basis of plant polyketide biosynthesis. *Nature Structural Biology* **6**, 775–784.

Fisher, R. F. and Long, S. R. (1992). *Rhizobium*-plant signal exchange. *Nature* **357**, 655–660.

Forkmann, G. (1977). Precursors and genetic control of anthocyanin synthesis in *Matthiola incana* R. Br. *Planta* **137**, 159–163.

Forkmann, G. and Martens, S. (2001). Metabolic engineering and applications of flavonoids. *Current Opinion in Biotechnology* **12**, 155–160.

Fritsch, H. and Grisebach, H. (1975). Biosynthesis of cyanidin in cell cultures of *Haplopappus gracilis*. *Phytochemistry* **14**, 2437–2442.

Geissman, T. A., Hinreiner, E. H. and Jorgensen, E. C. (1956). Inheritance in the carnation, *Dianthus caryophyllus*. V. The chemistry of flower color variation. II. *Genetics* **41**, 93.

Gough, C., Galera, C., Vasse, J., Webster, G., Cocking, E. C. and Dénarié, J. (1997). Specific flavonoids promote intercellular root colonization of *Arabidopsis thaliana* by *Azorhizobium caulinaodans* ORS571. *Molecular Plant–Microbe Interactions* **10**, 560–570.

Grotewold, E., Chamberlin, M., Snook, M., Siame, B., Butler, L., Swenson, J., Maddock, S., St. Clair, G. and Bowen, B. (1998). Engineering secondary metabolism in maize cells by ectopic expression of transcription factors. *The Plant Cell* **10**, 721–740.

Gruber, M. Y., Ray, H., Auser, P., Skadhauge, B., Falk, J., Thomsen, K. K., Stougaard, J., Muir, A., Lees, G., McKersie, B., Bowly, S. and von Wettstein, D. (1999). Genetic systems for condensed tannin biotechnology. *In* "Plant Polyphenols 2: Chemistry and Biology" (G. G. Gross, R. W. Hemingway and T. Yoshida, eds), pp. 315–341. Plenum Press, New York.

Harborne, J. B. and Williams, C. A. (2000). Advances in flavonoid research since 1992. *Phytochemistry* **55**, 481–504.

Harrison, M. J. and Dixon, R. A. (1994). Spatial patterns of expression of flavonoid/isoflavonoid pathway genes during interactions between roots of *Medicago truncatula* and the mycorrhizal fungus *Glomus versiforme*. *The Plant Journal* **6**, 9–20.

He, X.-Z. and Dixon, R. A. (2000). Genetic manipulation of isoflavone 7-*O*-methyltransferase enhances biosynthesis of 4′-*O*-methylated isoflavonoid phytoalexins and disease resistance in alfalfa. *The Plant Cell* **12**, 1689–1702.

Holton, T. A. and Cornish, E. C. (1995). Genetics and biochemistry of anthocyanin biosynthesis. *Plant Cell* **7**, 1071–1083.

Hrazdina, G. and Wagner, G. J. (1985a). Compartmentation of plant phenolic compounds; sites of synthesis and accumulation. *Annual Proceedings of the Phytochemical Society, Europe* **25**, 120–133.

Hrazdina, G. and Wagner, G. J. (1985b). Metabolic pathways as enzyme complexes: evidence for the synthesis of phenylpropanoids and flavonoids on membrane associated enzyme complexes. *Archives of Biochemistry and Biophysics* **237**, 88–100.

Hrazdina, G. and Jensen, R. A. (1992). Spatial organization of enzymes in plant metabolic pathways. *Annual Review of Plant Physiology and Plant Molecular Biology* **43**, 241–267.

Hrazdina, G., Zobel, A. M. and Hoch, H. C. (1987). Biochemical, immunological and immunocytochemical evidence for the association of chalcone synthase with endoplasmic reticulum membranes. *Proceedings of the National Academy of Sciences, USA* **84**, 8966–8970.

Jackson, D., Roberts, K. and Martin, C. (1992). Temporal and spatial control of expression of anthocyanin biosynthetic genes in developing flowers of *Antirrhinum majus*. *The Plant Journal* **2**, 425–434.

Jacobs, M. and Rubery, P. H. (1988). Naturally occurring auxin transport regulators. *Science* **241**, 346–349.

Jacques, D., Opie, C. T., Porter, L. J. and Haslam, E. (1977). Plant proanthocyanidins. Part 4. Biosynthesis of procyanidins and observations on the metabolism of cyanidin in plants. *Journal of the Chemical Society – Perkin Transactions I*, 1637–1643.

Jaffe, M. J. and Galston, A. W. (1967). Physiological studies on pea tendrils. IV. Flavonoids and contact coiling. *Plant Physiology* **23**, 848–850.

Jende-Strid, B. (1991). Gene-enzyme relations in the pathway of flavonoid biosynthesis in barley. *Theoretical and Applied Genetics* **81**, 668–674.

Jez, J. M., Austin, M. B., Ferrer, J.-L., Bowman, M. E., Schröder, J. and Noel, J. P. (2000). Structural control of polyketide formation in plant-specific polyketide synthases. *Chemical Biology* **7**, 919–930.

Jung, W., Yu, O., Sze-Mei, C. L., O'Keefe, D. P., Odell, J., Fader, G. and McGonigle, B. (2000). Identification and expression of isoflavone synthase, the key enzyme for biosynthesis of isoflavones in legumes. *Nature Biotechnology* **18**, 208–213.

Kappert, H. (1949). Die Genetik des *incana*-Charakters und der Anthozyanbildung bei der Levkoje. *Der Züchter* **19**, 289–297.

Kelly, C. K. (1990). Plant foraging: a marginal value model and coiling response in *Cuscuta subinclusa*. *Ecology* **71**, 1916–1925.

Kootstra, A. (1994). Protection from UV-B-induced DNA damage by flavonoids. *Plant Molecular Biology* **26**, 771–774.

Kroon, J., Souer, E., de Graaff, A., Xue, Y., Mol, J. and Koes, R. (1994). Cloning and structural analysis of the anthocyanin pigmentation locus Rt of *Petunia hybrida:* characterization of insertion sequences in two mutant alleles. *The Plant Journal* **5**, 69–80.

Landry, L. G., Chapple, C. C. S. and Last, R. L. (1995). Arabidopsis mutants lacking phenolic sunscreens exhibit enhanced ultraviolet-B injury and oxidative damage. *Plant Physiology* **109**, 1159–1166.

Leweke, B. and Forkmann, G. (1982). Genetically controlled anthocyanin synthesis in cell cultures of *Matthiola incana*. *Plant Cell Reports* **1**, 98–100.

Li, J., Ou-Lee, T.-M., Raba, R., Amundson, R. G. and Last, R. L. (1993). Arabidopsis flavonoid mutants are hypersensitive to UV-B irradiation. *Plant Cell* **5**, 171–179.

Loake, G. J., Choudhary, A. D., Harrison, M. J., Mavandad, M., Lamb, C. J. and Dixon, R. A. (1991). Phenylpropanoid pathway intermediates regulate transient expression of a chalcone synthase gene promoter. *The Plant Cell* **3**, 829–840.

Loake, G. J., Faktor, O., Lamb, C. J. and Dixon, R. A. (1992). Combination of H-box [CCTACC(N)7CT] and G-box (CACGTG) cis elements is necessary for feed-forward stimulation of a chalcone synthase promoter by the phenylpropanoid-pathway intermediate p-coumaric acid. *Proceedings of the National Academy of Sciences, USA* **89**, 9230–9234.

Margna, U. and Vainjarv, T. (1981). Buckwheat seedling flavonoids do not undergo rapid turnover. *Biochemie und Physiologie der Pflanzen* **176**, 44–53.

Martin, C., Prescott, A., Mackay, S., Bartlett, J. and Vrijlandt, E. (1991). Control of anthocyanin biosynthesis in flowers of *Antirrhinum majus*. *The Plant Journal* **1**, 37–49.

Mavandad, M., Edwards, R., Liang, X., Lamb, C. J. and Dixon, R. A. (1990). Effects of *trans*-cinnamic acid on expression of the bean phenylalanine ammonia-lyase gene family. *Plant Physiology* **94**, 671–680.

McKhann, H. I., Paiva, N. L., Dixon, R. A. and Hirsch, A. M. (1998). Expression of genes for enzymes of the flavonoid biosynthetic pathway in the early stages of the *Rhizobium*-legume symbiosis. *Advances in Experimental Medicine and Biology* **439**, 45–54.

Mo, Y., Nagel, C. and Taylor, L. P. (1992). Biochemical complementation of chalcone synthase mutants defines a role for flavonols in functional pollen. *Proceedings of the National Academy of Sciences, USA* **89**, 7213–7217.

Mol, J., Grotewold, E. and Koes, R. (1998). How genes paint flowers and seeds. *Trends in Plant Science* **3**, 212–217.

Mol, J., Cornish, E., Mason, J. and Koes, R. (1999). Novel coloured flowers. *Current Opinion in Biotechnology* **10**, 198–201.

Morris, P. and Robbins, M. P. (1997). Manipulating condensed tannins in forage legumes. *In* "Biotechnology and the Improvement of Forage Legumes" (B. D. McKersie and D. C. W. Brown, eds), pp. 147–173. CAB International, Wallingford, UK.

Muir, S. R., Collins, G. J., Robinson, S., Hughes, S., Bovy, A., De Vos, C. H. R., van Tunen, A. J. and Verhoeyen, M. E. (2001). Overexpression of petunia chalcone isomerase in tomato results in fruit containing increased levels of flavonols. *Nature Biotechnology* **19**, 470–474.

Müller, A., Guan, C., Gälweiler, L., Tänzler, P., Huijser, P., Marchant, A., Parry, G., Bennett, M., Wisman, E. and Palme, K. (1998). *AtPIN2* defines a locus of *Arabidopsis* for root gravitropism control. *The EMBO Journal* **17**, 6903–6911.

Murphy, A., Peer, W. A. and Taiz, L. (2000). Regulation of auxin transport by aminopeptidases and endogenous flavonoids. *Planta* **211**, 315–324.

Napoli, C. A., Fahy, D., Wang, H. Y. and Taylor, L. P. (1999). *White anther:* A petunia mutant that abolishes pollen flavonol accumulation, induces male sterility, and is complemented by a chalcone synthase transgene. *Plant Physiology* **120**, 615–622.

Nesi, N., Debeaujon, I., Jond, C., Pelletier, G., Caboche, M. and Lepiniec, L. (2000). The *TT8* gene encodes a basic helix-loop-helix domain protein required for expression of *DFR* and *BAN* genes in Arabidopsis siliques. *The Plant Cell* **12**, 1863–1878.

Ormrod, D. P., Landry, L. G. and Conklin, P. L. (1995). Short-term UV-B radiation and ozone exposure effects on aromatic secondary metabolite accumulation and shoot growth in flavonoid-deficient *Arabidopsis* mutants. *Physiologia Plantarum* **93**, 602–610.

Ovádi, J., and Srere, P. A. (2000). Macromolecular compartmentation and channeling. *International Review of Cytology* **192**, 255–280.

Pelletier, M. K. and Shirley, B. W. (1995). A genomic clone encoding flavanone 3-hydroxylase (accession no. U33932) from *Arabidopsis thaliana. Plant Physiology* **109**, 1125–1127.

Pelletier, M. K., Murrell, J. and Shirley, B. W. (1997). Arabidopsis flavonol synthase and leucoanthocyanidin dioxygenase: further evidence for distinct regulation of 'early' and 'late' flavonoid biosynthetic genes. *Plant Physiology* **113**, 1437–1445.

Pelletier, M. K., Burbulis, I. E. and Shirley, B. W. (1999). Disruption of specific flavonoid genes enhances the accumulation of flavonoid enzymes and endproducts in *Arabidopsis* seedlings. *Plant Molecular Biology* **40**, 45–54.

Pollak, P. E., Vogt, T., Mo, Y. and Taylor, L. P. (1993). Chalcone synthase and flavonol accumulation in stigmas and anthers of *Petunia hybrida. Plant Physiology* **102**, 925–932.

Quattrocchio, F., Wing, J. F., Leppen, H. T. C., Mol, J. N. M. and Koes, R. E. (1993). Regulatory genes controlling anthocyanin pigmentation are functionally conserved among plant species and have distinct sets of target genes. *The Plant Cell* **5**, 1497–1512.

Rasmussen, S. and Dixon, R. A. (1999). Transgene-mediated and elicitor-induced perturbation of metabolic channeling at the entry point into the phenylpropanoid pathway. *The Plant Cell* **11**, 1537–1551.

Rice-Evans, C. (2001). Flavonoid antioxidants. *Current Medicinal Chemistry* **8**, 797–807.

Saslowsky, D. and Winkel-Shirley, B. (2001). Localization of flavonoid enzymes in *Arabidopsis* roots. *The Plant Journal* **27**, 37–48.

Saslowsky, D. E., Dana, C. D. and Winkel-Shirley, B. (2000). An allelic series for the chalcone synthase locus in Arabidopsis. *Gene* **255**, 127–138.

Selinger, D. A. and Chandler, V. L. (1999). A mutation in the *pale aleurone color1* gene identifies a novel regulator of the maize anthocyanin pathway. *The Plant Cell* **11**, 5–14.

Shirley, B. W. and Hwang, I. (1995). The interaction trap: *In vivo* analysis of protein-protein associations. *In* "Methods in Plant Cell Biology, Part A" (D. W. Galbraith, H. J. Bohnert and D. P. Bourque, eds), Vol. 49, pp. 401–416. Academic Press, San Diego.

Shirley, B. W., Kubasek, W. L., Storz, G., Bruggemann, E., Koornneef, M., Ausubel, F. M. and Goodman, H. M. (1995). Analysis of *Arabidopsis* mutants deficient in flavonoid biosynthesis. *The Plant Journal* **8**, 659–671.

Stafford, H. A. (1974). Possible multi-enzyme complexes regulating the formation of C_6-C_3 phenolic compounds and lignins in higher plants. *Recent Advances in Phytochemistry* **8**, 53–79.

Stafford, H. A. (1981). Compartmentation in natural product biosynthesis by multienzyme complexes. *In* "The Biochemistry of Plants" (E. E. Conn, ed.), Vol. 7, pp. 117–137. Academic Press, New York.

Stafford, H. A. (1991). Flavonoid evolution: an enzymic approach. *Plant Physiology* **96**, 680–685.

Stapleton, A. E. and Walbot, V. (1994). Flavonoids can protect maize DNA form the induction of ultraviolet radiation damage. *Plant Physiology* **105**, 881–889.

Tanaka, Y., Tsuda, S. and Kusumi, T. (1998). Metabolic engineering to modify flower color. *Plant Cell Physiology* **39**, 1119–1126.

Taylor, L. P. and Jorgensen, R. (1992). Conditional male fertility in chalcone synthase-deficient petunia. *Journal of Heredity* **83**, 11–17.

van der Meer, I. M., Stam, M. E., van Tunen, A. J., Mol, J. N. M. and Stuitje, A. R. (1992). Antisense inhibition of flavonoid biosynthesis in petunia anthers results in male sterility. *The Plant Cell* **4**, 253–262.

van Tunen, A. J., Mur, L. A., Recourt, K., Gerats, A. G. M. and Mol, J. N. M. (1991). Regulation and manipulation of flavonoid gene expression in anthers of petunia: the molecular basis of the *Po* mutation. *The Plant Cell* **3**, 39–48.

von Wettstein, D., Jende-Strid, B., Ahrenst-Larsen, B. and Erdal, K. (1980). Proanthocyanidin-free barley prevents the formation of beer haze [Analysis of varieties and mutants]. *Technical Quarterly of the Master Brewers Association of the America* **17**, 16–23.

Wagner, G. J. and Hrazdina, G. (1984). Endoplasmic reticulum as a site of phenylpropanoid and flavonoid metabolism in *Hippeastrum*. *Plant Physiology* **74**, 901–906.

Weiss, M. R. (1991). Floral colour changes as cues for pollinators. *Nature* **354**, 227–229.

Western, T. L., Burn, J., Tan, W. L., Skinner, D. J., Martin-McCaffrey, L., Moffatt, B. A. and Haughn , G. W. (2001) Isolation and characterization of mutants defective in seed coat mucilage secretory cell development in *Arabidopsis*. *Plant Physiology* **127**, 998–1011.

Westwood, J. H. (2000). Characterization of the *Orobanche-Arabidopsis* system for studying parasite-host interactions. *Weed Science* **48**, 742–748.

Willson, M. F. and Whelen, C. J. (1990). The evolution of fruit color in fleshy-fruited plants. *American Naturalist* **136**, 790–809.

Winkel-Shirley, B. (2001a). Flavonoid biosynthesis: a colorful model for genetics, biochemistry, cell biology and biotechnology. *Plant Physiology* **126**, 485–493.

Winkel-Shirley, B. (2001b). It takes a garden. How work on diverse plant species has contributed to an understanding of flavonoid metabolism. *Plant Physiology* **127**, in press.

Ylstra, B., Touraev, A., Moreno, R. M. B., Stöger, E., van Tunen, A. J., Vicente, O., Mol, J. N. M. and Heberle-Bors, E. (1992). Flavonols stimulate development, germination, and tube growth of tobacco pollen. *Plant Physiology* **100**, 902–907.

Ylstra, B., Busscher, J., Franken, J., Hollman, P. C. H., Mol, J. N. M. and van Tunen, A. J. (1994). Flavonols and fertilization in *Petunia hybrida:* localization and mode of action during pollen tube growth. *The Plant Journal* **6**, 201–212.

Ylstra, B., Muskens, M. and van Tunen, A. J. (1996). Flavonols are not essential for fertilization in *Arabidopsis thaliana*. *Plant Molecular Biology* **32**, 1155–1158.

Yu, O., Jung, W., Shi, J., Croes, R. A., Fader, G. M., McGonigle, B. and Odell, J. T. (2000). Production of the isoflavones genistein and daidzein in non-legume dicot and monocot tissues. *Plant Physiology* **124**, 781–793.

Zheng, D., Schröder, G., Schröder, J. and Hrazdina, G. (2001). Molecular and biochemical characterization of three aromatic polyketide synthase genes from *Rubus idaeus*. *Plant Molecular Biology* **46**, 1–15.

Differential Expression and Functional Significance of Anthocyanins in Relation to Phasic Development in *Hedera helix* L.

WESLEY P. HACKETT

Department of Environmental Horticulture, University of California, Davis, CA 95616, USA

ABSTRACT

During the development of plants from seed, there are changes in their phenotypic characteristics. These changes in characteristics are referred to as maturation, phase change or phasic development. Those characteristics that appear early in development are referred to as juvenile and those that appear late are referred to as mature or adult. The characteristics that change during phasic development are not necessarily consistent from species to species. However, phase specific accumulation of anthocyanins in stem and leaf tissues has been observed in a number of genera and species. This paper presents experimental evidence for the anatomical, biochemical, and molecular bases for phase specific accumulation of anthocyanin in *Hedera helix,* a temperate zone vine species. The relationship of phase specific anthocyanin accumulation with other physiological and anatomical characteristics and the possible functional significance will be discussed. Because of the demonstrated anthocyanin accumulation response of juvenile, but not mature phase, leaf discs to high light and increased sugar concentrations, it is proposed that anthocyanin has a photoprotective function in the shade adapted juvenile leaves which have a lesser developed capability to acclimate to high light conditions than mature phase leaves.

Advances in Botanical Research Vol. 37
incorporating Advances in Plant Pathology
ISBN 0-12-005937-1

I. INTRODUCTION

Phasic development has been intriguing but baffling to applied and basic scientists for many years (Goebel, 1900). The morphological and physiological characteristics associated with phase change have been described for many species, especially woody perennial species (Hackett, 1985). However, little is known about the mechanisms involved in the change of characteristics or the functional significance of the changes. The main objective of this chapter is to illustrate experimental approaches that have contributed to a better understanding of the anatomical, biochemical, and molecular bases for phase specific accumulation of anthocyanin in *Hedera helix.* In addition the functional significance of anthocyanin accumulation in relation to other phase specific characteristics in *Hedera helix* will be discussed.

Hedera helix L. (English ivy) is a classical example of phasic development in which characteristics of the juvenile and mature phases are quite distinct (Table I). It is a temperate zone vine species that grows naturally on the shaded forest floor. Because of its stem aerial roots, it can cling to and grow up the trunks of trees. When growing on the forest floor it has juvenile characteristics. However, as it grows up the trunks of trees into a different environment, its vegetative characteristics gradually change (Table I) and it acquires the ability to flower after a poorly defined time period (Wareing and Frydman, 1976).

II. CONCEPT AND CHARACTERISTICS OF PHASIC DEVELOPMENT

During the development of plants from seed there are changes in the characteristics of the structures produced by the shoot apical meristem. The most obvious of these is the transition from vegetative to reproductive structures. This change is a reflection of a change in competence to perceive, or respond to environmental or endogenous stimuli for flower

TABLE I
Some characteristics that distinguish the juvenile and mature phases of Hedera helix

Characteristic	Juvenile	Mature
Shoot apex width	*c.* 140 μm	*c.* 200 μm
Plastochron	4.2 days	3.2 days
Internode length	Long	Short
Leaf shape	Five lobed, palmate	Entire, ovate
Phyllotaxis	Distichous (1, 2)	Spiral (2, 3)
Growth habit	Plagiotropic	Orthotropic
Anthocyanin stem pigmentation	Present	Absent
Photosynthetic capability	Low	High
Stem aerial roots	Present	Absent
Ability to flower	Absent	Present

induction. In addition, there are progressive changes in vegetative characteristics that involve morphological, anatomical, physiological, and developmental differences. These include leaf shape, thickness, and epidermal characteristics, phyllotaxis, thorniness, shoot orientation, shoot growth vigor, anthocyanin, photosynthetic characteristics, disease and insect resistance, and competence to form adventitious buds and roots and somatic embryos (Hackett, 1985). Degree of change in such characteristics during development varies from species to species and most change gradually resulting in intermediate or transitional forms and competencies (Hackett, 1985). In addition, the various characteristics may change at different rates in a given species (Greenwood *et al.*, 1989). Those characteristics that occur early in development are referred to as juvenile and those that appear late are referred to as mature or adult characteristics. The change from juvenile to mature characteristics is referred to as maturation or phase change. Phasic differences in characteristics are most obvious in woody perennial species because of their prolonged juvenile and mature phases (Table I) but are also observed in herbaceous species like *Zea mays* and *Arabidopsis thaliana* (Poethig, 1990).

III. ANTHOCYANIN ACCUMULATION IN RELATION TO PHASIC DEVELOPMENT

An extensive list of species with differential accumulation of anthocyanin in juvenile and mature phases has not been compiled. However, there are sufficient references to juvenile phase specific accumulation of anthocyanins in stem and leaf tissues in woody perennial genera, including *Betula* (Brand and Linberger, 1992), *Carya* (Romberg, 1944), *Malus* (Stoutemyer, 1937), *Pinus* (Nozzolillo *et al.*, 1990; Camm *et al.*, 1993), and *Rhododendron* (Balfour, 1919), to indicate that phase specific accumulation is relatively common. In addition, there are many references to anthocyanin accumulation in newly germinated seedlings of herbaceous species that disappears days or weeks after germination (Chalker-Scott, 1999). There are also references to juvenile leaf reddening, particularly in tropical species, in which anthocyanins accumulate in expanding young leaves but disappear as the leaves become fully expanded. Based on literature descriptions (Opler *et al.*, 1980), this phenomenon appears to occur both in leaves of seedlings and fully developed (mature) plants and therefore, such accumulation of anthocyanin should not be considered to be related to developmental phase.

IV. PHASE-RELATED ANTHOCYANIN ACCUMULATION IN *HEDERA HELIX*

A. ORGAN AND TISSUE LOCALIZATION

Juvenile phase *H. helix* accumulates anthocyanin (cyanidin) in the hypodermis of stems and leaf petioles but not in the leaf lamina when grown at temperatures above 15°C. In the hypodermis of juvenile stems and petioles, it is the thick walled collenchyma cell layer that accumulates anthocyanin. Under the appropriate temperature conditions, juvenile leaf lamina accumulate anthocyanin in the palisade cells of the mesophyll. Mature phase ivy does not accumulate anthocyanin in anatomically similar hypodermal tissue or leaf lamina when grown under the same temperature conditions.

B. INDUCTION OF ANTHOCYANIN ACCUMULATION IN JUVENILE LAMINA TISSUE

As indicated above, neither juvenile and mature lamina tissues accumulate anthocyanin at temperatures above 15°C. Murray and Hackett (1991) used excised lamina discs from juvenile and mature phase leaves, incubated in petri dishes, to study factors that influence accumulation of anthocyanin. In experiments in which light levels and CO_2 concentrations were varied, both juvenile and mature phase lamina discs accumulated methanol extractable phenylpropanoids in response to light, with greater accumulation at increased CO_2 concentrations. However, only juvenile phase discs accumulated anthocyanin in response to light and CO_2. Greater concentrations of anthocyanin accumulated in juvenile discs in response to increased photosynthetic photon flux (PPFD) and increased CO_2 concentrations. The dependence of anthocyanin accumulation in juvenile lamina discs on both PPFD level and CO_2 concentration indicates that the major effect of light is a result of increased photosynthetic production of carbohydrates. The importance of carbohydrates in induction of anthocyanin in juvenile phase lamina tissue was confirmed based on results which showed that anthocyanins accumulate under low CO_2 levels when supplemented with exogenously supplied carbohydrates. Sucrose, glucose and fructose but not rhamnose, a 6-deoxyhexose, were effective in inducing anthocyanin. The lack of response to rhamnose indicates that the positive response to sugars is not a response to osmotic potential in treatment solutions. Juvenile phase lamina discs treated with sucrose in the dark did not accumulate anthocyanin indicating a photomorphogenetic as well as a photosynthetic response to light.

C. BASIS OF DIFFERENTIAL ANTHOCYANIN ACCUMULATION IN JUVENILE AND MATURE PHASE TISSUES

1. *Biochemical*

As indicated in the previous section, both juvenile and mature phase leaf discs accumulate methanol extractable phenylpropanoids under conditions favorable for anthocyanin accumulation in juvenile discs. This indicates that the lack of accumulation of anthocyanin in mature phase tissue is not due to a limitation of phenylpropanoid precursors of anthocyanins. In addition, when leaf discs were incubated on sucrose in light, Murray and Hackett (1991) found that lamina discs of both phases accumulated comparable levels of the flavanols kaemferol and quercetin, whereas only juvenile phase discs accumulated anthocyanin. The accumulation of comparable levels of flavanols in both phases suggests that enzymes early in the flavanoid pathway, such as chalcone synthase (CHS) are not limiting in mature phase tissues. The accumulation of quercetin, but lack of accumulation of leucoanthocyanin or anthocyanin in mature phase discs, suggests that the mature phase tissue lacks dihydroflavanol reductase (DFR) activity. Subsequent enzyme assays showed there was no detectable DFR activity in mature phase lamina discs treated with light and sucrose, whereas there was induction of DFR activity in juvenile phase discs given the same treatment. Murray and Hackett (1991) conclude that the lack of DFR activity limits the accumulation of anthocyanin in mature phase tissue.

2. *Molecular*

Based on the work described in the previous section, Murray *et al.* (1994) determined the level of gene expression that is associated with the differential DFR activity in juvenile and mature phase lamina tissue of ivy. To do this, DNA clones of DFR and CHS were isolated from ivy using DFR and CHS DNA clones from other species to probe an ivy genomic library. By using nuclear run-on transcription assays and RNA blot analysis, it was demonstrated that the lack of induction of DFR activity by light and sucrose in mature phase lamina tissue is due to the lack of transcription of the DFR gene and the resultant lack of accumulation of detectable DFR mRNA. In contrast, lamina tissue of both phases transcribe the CHS gene and accumulate CHS mRNA in response to light and sucrose indicating mature phase tissue is responsive to the induction treatment. These responses are not unique to lamina tissues in that CHS mRNA accumulates in both juvenile and mature phase stem tissues, but DFR mRNA accumulates only in juvenile phase stem tissue. Murray *et al.* (1994) concluded that the lack of competence for induction of anthocyanin pigmentation in lamina tissue of mature phase ivy is due to the lack of transcription of the DFR gene.

V. FUNCTIONAL SIGNIFICANCE OF ANTHOCYANIN ACCUMULATION IN *HEDERA HELIX*

Bauer and Bauer (1980) have studied the photosynthetic and anatomical characteristics of leaves from juvenile and mature phase *H. helix* in great detail. They concluded that these characteristics of leaves of juvenile and mature phase plants resemble those reported for leaves of shade and sun adapted plants, respectively, even though the ivy leaves studied originated from plants grown in the same light environment. Hoflacher and Bauer (1982) and Bauer and Thoni (1988) concluded that leaves of both the juvenile and mature phases have the capability to acclimate to strong light. However, leaves of the juvenile phase have less absolute capacity to acclimate than those of the mature phase.

Woo *et al.* (1994) showed that steady state levels of mRNA which encodes an LHCII protein (light harvesting chlorophyll a/b binding protein of photosystem II) are higher in juvenile than mature phase lamina from plants grown under the same environmental conditions. They speculate that these differences in *Lhcb* gene expression may be related to differences in photosynthetic and acclimation characteristics and described by Bauer and his co-workers (see above) as being similar to differences in leaves of sun and shade adapted plants. It is known that the proportion of lamellae-forming grana and the ratio of thylakoid membranes to stroma is generally greater in shade than sun adapted plants (Anderson *et al.*, 1988). Because juvenile phase ivy is a shade adapted plant, it may possess chloroplast structure with increased stacking of the thylakoid membranes in photosystem II (PSII) for efficient capture of low irradiance, far-red enriched light energy in the shade habitat. Since a major function of LHCII proteins is to provide adhesion between thylakoid membranes (Staehelin and Arntzen, 1983), the higher expression of the *Lhcb* gene in juvenile phase ivy leaves may reflect the development and maintenance of a larger LHCII complex in juvenile than mature phase leaf chloroplasts even when exposed to the same irradiance levels. Unpublished data (Woo, H-H. and Hackett, W. P.) show that leaves of juvenile phase ivy have high levels of expression of the *Lhcb* gene whether developed in high or low irradiance light whereas leaves of the mature phase have reduced expression when developed in high as compared to low irradiance. This suggests that mature phase leaves have greater capability to acclimate than juvenile ones. These data are consistent with the acclimation findings of Bauer and co-workers for ivy cited above.

Since juvenile phase *H. helix* leaves have a lesser ability to acclimate to high light levels as indicated above, canopy gaps formed by tree blowdown could result in photoinhibition or oxidative damage to the photosynthetic apparatus, especially at low temperature, unless juvenile shade adapted leaves have some photoprotective mechanism. Because of the demonstrated anthocyanin accumulation response of juvenile phase ivy lamina discs to high light and increased sucrose concentration, it is proposed that anthocyanin has a photoprotective function in the shade adapted juvenile phase leaves. Such a role for anthocyanin has been suggested by Kroll *et al.* (1995) for seedlings of jack pine, *Pinus banksiana*. This hypothesis is testable for *H. helix* since anthocyanin accumulation is environmentally inducible in juvenile but not mature phase leaves by manipulating CO_2, light, and temperature (less than 15°C).

REFERENCES

Anderson, J., Chow, W. S. and Goodchild, D. J. (1988). Thylakoid membrane organization in sun/shade acclimation. *Australian Journal of Plant Physiology* **15**, 11–16.

Balfour, B. (1919). Observations on rhododendron seedlings. *Transactions of the Botanical Society of Edinburgh* **27**, 366–372.

Bauer, H. and Bauer, U. (1980). Photosynthesis in leaves of the juvenile and adult phase of ivy (*Hedera helix*). *Physiologia Plantarum* **49**, 366–372.

Bauer, H. and Thoni, W. (1988). Photosynthetic light acclimation in fully developed leaves of the juvenile and adult life phases of *Hedera helix*. *Physiologia Plantarum* **73**, 31–37.

Brand, M. H. and Lineberger, L. D. (1992). *In vitro* rejuvenation of *Betula* (*Betulaceae*): morphological evaluation. *American Journal of Botany* **79**, 618–625.

Camm, E. L., McCallum, J., Leaf, E. and Koupai-Abyazani, M. R. (1993). Cold-induced purpling of *Pinus contorta* seedlings depends on previous daylength treatment. *Plant, Cell and Environment* **16**, 761–764.

Chalker-Scott, L. (1999). Environmental significance of anthocyanins in plant stress responses. *Photochemistry and Photobiology* **70**, 1–9.

Goebel, K. (1900). "Organography of Plants". Clarendon Press, Oxford.

Greenwood, M. S., Hooper, C. A. and Hutchison, K. W. (1989). Maturation in larch. I. Effects of age on shoot growth, foliar characteristics, and DNA methylation. *Plant Physiology* **90**, 406–412.

Hackett, W. P. (1985). Juvenility, maturation and rejuvenation in woody plants. *Horticultural Reviews* **7**, 109–155.

Hoflacher, H. and Bauer, H. (1982). Light acclimation in leaves of the juvenile and adult phases of ivy (*Hedera helix*). *Physiologia Plantarum* **56**, 177–182.

Kroll, M., Gray, G. R., Hurry, M. V., Oquist, G. and Huner, N. P. A. (1995). Low-temperature stress and photoperiod affect an increased tolerance to photoinhibition in *Pinus banksiana* seedlings. *Canadian Journal of Botany* **73**, 1119–1127.

Murray, J. R. and Hackett, W. P. (1991). Dihydroflavanol reductase activity in relation to differential anthocyanin accumulation in juvenile and mature phase *Hedera helix* L. *Plant Physiology* **97**, 343–351.

Murray, J. R., Smith, A. G. and Hackett, W. P. (1994). Differential dihydrolflavanol reductase transcription and anthocyanin accumulation in juvenile and mature phases of ivy (*Hedera helix* L.). *Planta* **194**, 102–107.

Nozzolillo, C. Isabelle, P. and Das, G. (1990). Seasonal changes in the phenolic constituents of jack pine seedlings in relation to the purpling phenomenon. *Canadian Journal of Botany* **68**, 2017–2021.

Opler, P. A., Frankie, G. W. and Baker, G. H. (1980). Comparative phenological studies of treelet and shrub species in wet and dry forest in the lowlands of Costa Rica. *Journal of Ecology* **68**, 167–188.

Poethig, R. S. (1990). Phase change and regulation of shoot morphogenesis in plants. *Science* **34**, 923–930.

Romberg, L. D. (1944). Some characteristics of the juvenile and the bearing pecan tree. *Proceedings of the American Society for Horticultural Science* **44**, 255–259.

Staehelin, L. A. and Arntzen, C. J. (1983) Regulation of chloroplast membrane function: Protein phosphorylation changes the spatial organization of membrane components. *Journal of Cell Biology* **97**, 1327–1337.

Stoutemyer, V. T. (1937). Regeneration in various types of apple wood. *Research Bulletin of the lowa Agricultural Experiment Station* **220**, 308–352.

Wareing, P. F. and Frydman, V. M. (1976). General aspects of phase change with special reference to *Hedera helix* L. *Acta Horticulturae* **56**, 57–68.

Woo, H-H., Hackett, W. P. and Das, A. (1994). Differential expression of a chlorophyll a/b binding protein gene and a proline rich protein gene in juvenile and mature phase English ivy (*Hedera helix*). *Physiologia Plantarum* **92**, 69–78.

Do Anthocyanins Function as Osmoregulators in Leaf Tissues?

LINDA CHALKER-SCOTT

Division of Ecosystem Sciences, College of Forest Resources, Box 354115, University of Washington, Seattle, WA 98195, USA

ABSTRACT

Water stress can be induced in plant tissues, directly or indirectly, by a number of environmental conditions. Plants exposed to drought, heat, cold, wind, flooding, or saline conditions often synthesize foliar anthocyanins in response. Although previously thought to function as a UV screen, anthocyanins may instead serve to decrease leaf osmotic potential. The resulting depression of leaf water potential could increase water uptake and/or reduce transpirational losses. Combined with other anti-stress activities attributed to anthocyanins (including their solar shield and antioxidative capacities), this phenomenon may allow anthocyanin-containing leaves to tolerate suboptimal water levels. The often transitory nature of foliar anthocyanin accumulation may allow plants to respond quickly and temporarily to environmental variability rather than through more permanent anatomical or morphological modifications.

Advances in Botanical Research Vol. 37
incorporating Advances in Plant Pathology
ISBN 0-12-005937-1

I. INTRODUCTION

Generally associated with the brilliant red, blue, and purple coloration seen in floral tissues, anthocyanins (along with the betacyanins) are also common foliar constituents in a diverse assortment of plants. Anthocyanins are water-soluble pigments derived from other flavonoids via the shikimic acid pathway. While foliar anthocyanins may be permanent leaf components in some species, they are developmentally or environmentally transient in many others. Perhaps best known are the transitory autumnal anthocyanins associated with deciduous leaf senescence in temperate regions. At the other end of the developmental spectrum are the juvenile anthocyanins, which appear at bud break and generally disappear after leaves are fully expanded. Finally, anthocyanins can be environmentally transient, appearing and disappearing with changes in photoperiod, temperature, and other signals. The relationship between transient anthocyanins and environmental stress resistance has been recently reviewed (Chalker-Scott, 1999).

Regardless of the developmental or environmental signal responsible, anthocyanic transcience can represent a significant metabolic cost to the plant. Energy is required to modify flavanone precursors and form anthocyanins *de novo*; likewise, their degradation also requires energy. A second potential cost of anthocyanin accumulation is the resulting interference with the light reactions of photosynthesis. Because of their ability to absorb blue and reflect red wavelengths, anthocyanins in the upper epidermis or mesophyll of leaves theoretically compete with light harvesting by chlorophyll and carotenoids. Reductions in photosynthetic rates have been noted in red-leafed varieties of *Coleus* (Burger and Edwards, 1996) and *Capsicum annuum* (Bahler *et al.*, 1991), spring flushing leaves of *Brachystegia* spp. (Tuohy and Choinski Jr, 1990; Choinski and Johnson, 1993), and the red juvenile leaves of several rainforest tree species (Woodall *et al.*, 1998). The competitive advantage afforded to these plants by accumulating anthocyanins must outweigh the costs associated with manufacturing and storing these compounds. This may especially be true under environmentally stressful conditions.

In this chapter the relationships between anthocyanins and abiotic environmental stressors will be explored at the whole plant level, particularly those stressors that directly or indirectly induce dehydration stress. These stressors include drought, osmotic disrupters, cold temperature, and anoxia. Evidence in the literature that coincidentally or deliberately links the presence of anthocyanins with induced resistance will be discussed. This concept of environmental resistance can then be applied to specific life history events in an attempt to explain why some anthocyanins are developmentally transient in the context of osmotic regulation.

II. INTERACTIONS AMONG ENVIRONMENTAL STRESSORS, PLANT WATER CONTENT, AND ANTHOCYANINS

A. DROUGHT STRESS

The term 'drought' is generally used in reference to an abnormal reduction in available water. Certain environments characterized as arid, therefore, are not droughty unless they receive a significantly less than normal amount of rainfall. For the purposes of this chapter, the phrase 'drought stress' will be used more generally to refer to any condition in which water may be a limiting factor.

1. Impact of Drought Stress upon Plant Water Relations

Lack of available water, whether or not a normal occurrence in a given environment, has the immediate effect of decreasing turgor, particularly in leaves. As evapotranspiration exceeds water uptake, loss of turgor generally triggers stomatal closure and a decrease in photosynthetic activity. Under prolonged stress, cellular dehydration and eventually plasmolysis will occur in drought-sensitive species. Plants that survive drought conditions generally have small, thick leaves with substantial cuticles, pubescence and/or other structural modifications (Curtis *et al.*, 1996; Li *et al.*, 1996). In contrast, actively growing tissues, whose cell walls are still expanding, often increase vacuolar solutes to decrease leaf osmotic potential, allowing the plant to remain turgid under low soil water potential conditions.

Though most water stress probably occurs as a result of low soil water availability, other environmental variables can induce water stress at the leaf level. Anything that causes a loss of integrity of the leaf cuticle and/or epidermis will induce increased evapotranspiration. Strong oxidizing agents, such as ozone, are able to breach these barriers and have been associated in the literature with leaf water stress (Gunthardt-Goerg *et al.*, 1993; Landolt *et al.*, 1994; Johnson *et al.*, 1995; Sakata, 1996; Maier-Maercker, 1997). Under such conditions, evapotranspiration exceeds water uptake and leaf water stress results.

2. Evidence of Anthocyanin-mediated Drought Resistance

a. Linkages to Morphological Adaptations to Drought. Both anecdotal and experimental observations have linked drought tolerance with foliar morphological characteristics such as thickened cell walls and/or cuticles, increased pubescence, and anthocyanin accumulation. Several decades ago, research on *Populus* spp. revealed positive correlations between petiole hairiness and anthocyanin content (Fritzsche and Kemmer, 1959; Kemmer and Fritzsche, 1961). Even earlier work on *Fraxinus americana* populations had determined that the southernmost ecotype exhibited both increased anthocyanin content and leaf pubescence (Wright, 1944); presumably, this ecotype would be exposed to hotter, drier conditions than more northern populations. Foliar anthocyanins have also been correlated with pubescence in *Fragaria* spp. (Sjulin *et al.*, 2000) and glaucousness in *Eucalyptus urnigera* (Thomas and Barber, 1974).

There is evidence that these adaptations to drought are more than coincidental. Recent molecular work with *Arabidopsis* mutants has linked these characteristics at the genetic level. Shirley *et al.* (1995) studied several loci responsible for foliar anthocyanin development and discovered that one of these was also responsible for trichome development. Thickened leaves and anthocyanin increases were found in an *Arabidopsis* mutant adapted to high light conditions (Iida *et al.*, 2000); high light levels are known to cause secondary drought stress. Finally, a population of *Arabidopsis* mutants with reduced cuticular waxes was found to accumulate more anthocyanins than normal (Millar *et al.*, 1998), presumably as a means to maintain foliar water.

b. Linkages to Ecophysiological Adaptations to Drought. Although not commonly mentioned in context with plants from arid environments, anthocyanins are nevertheless associated with water stress in a variety of studies. Perhaps the best example of anthocyanin-associated drought tolerance is that exhibited by resurrection plants. These plants, which include the genera *Craterostigma*, *Myrothamnus*, and *Xerophyta*, are able to lose most of their available water, yet remain viable for extended periods of time. During this process, resurrection plants replace their large central vacuole with smaller vacuoles and increase their anthocyanin content three- to fourfold over normal levels (Sherwin and Farrant, 1998; Farrant, 2000).

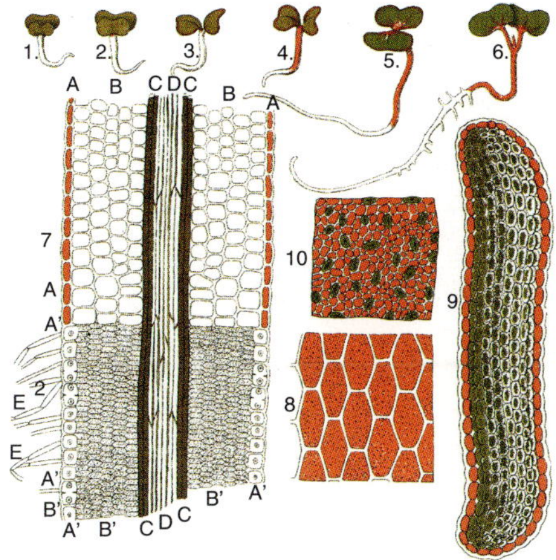

Plate 1. The distribution of anthocyanins in the cell sap of various organs of young seedlings of red cabbage, *Brassica oleracea*, from microscope observations by Édouard Morren (1858).

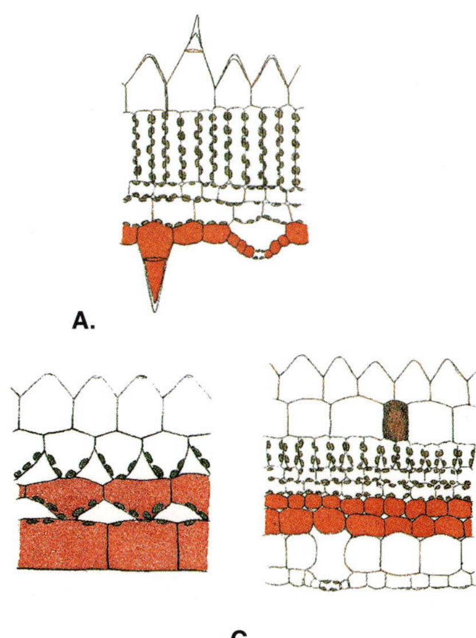

Plate 2. Distribution of anthocyanins in leaf tissues of tropical rainforest understory plants observed in Java (Stahl, 1896). A. *Eranthemum cooperi* (Acanthaceae), lower epidermis. B. *Piper porphyraceum* (Piperaceae), 'spongy' mesophyll. C. *Begonia falcifolia* (Begoniaceae), lower epidermis and single layer of spongy mesophyll.

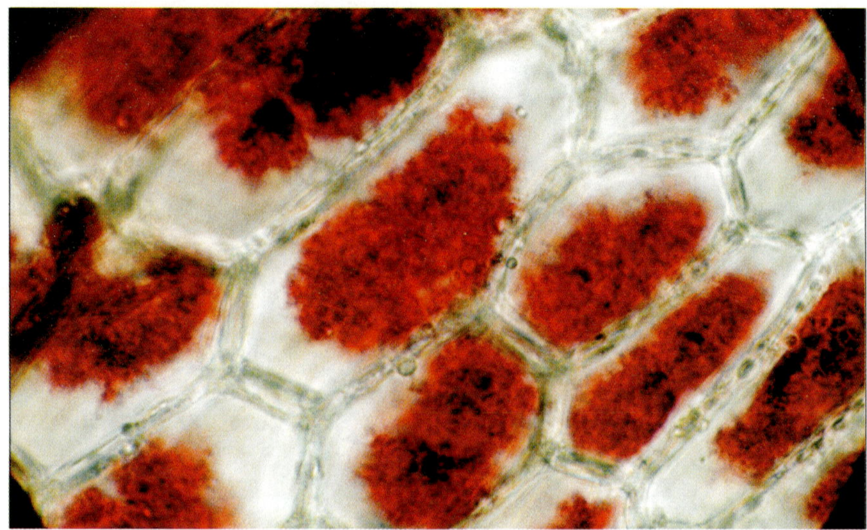

Plate 3. Anthocyanic vacuolar inclusions (AVIs) in epidermal peels taken from the inner petal region of flowers isolated from lisianthus (*Eustoma grandiflorum* cv. Excel Cherry) (×40).

Plate 4. View of anthocyanic young leaves in the ornamental shrub, *Photinia* sp., photograph on left of entire shrub, and that on right a detail of the young leaves.

Plate 5. *Ledum palustre* immediately after snowmelt (upper) and mid season (middle) showing changes in pigmentation and leaf orientation. Lower panel shows cross section (×40) of *Empetrum nigrum* revealing location of anthocyanin containing cells.

Plate 6. A collection of autumn leaves, both red and yellow, collected from the Harvard Forest in Central Massachusetts on 2 October 1998.

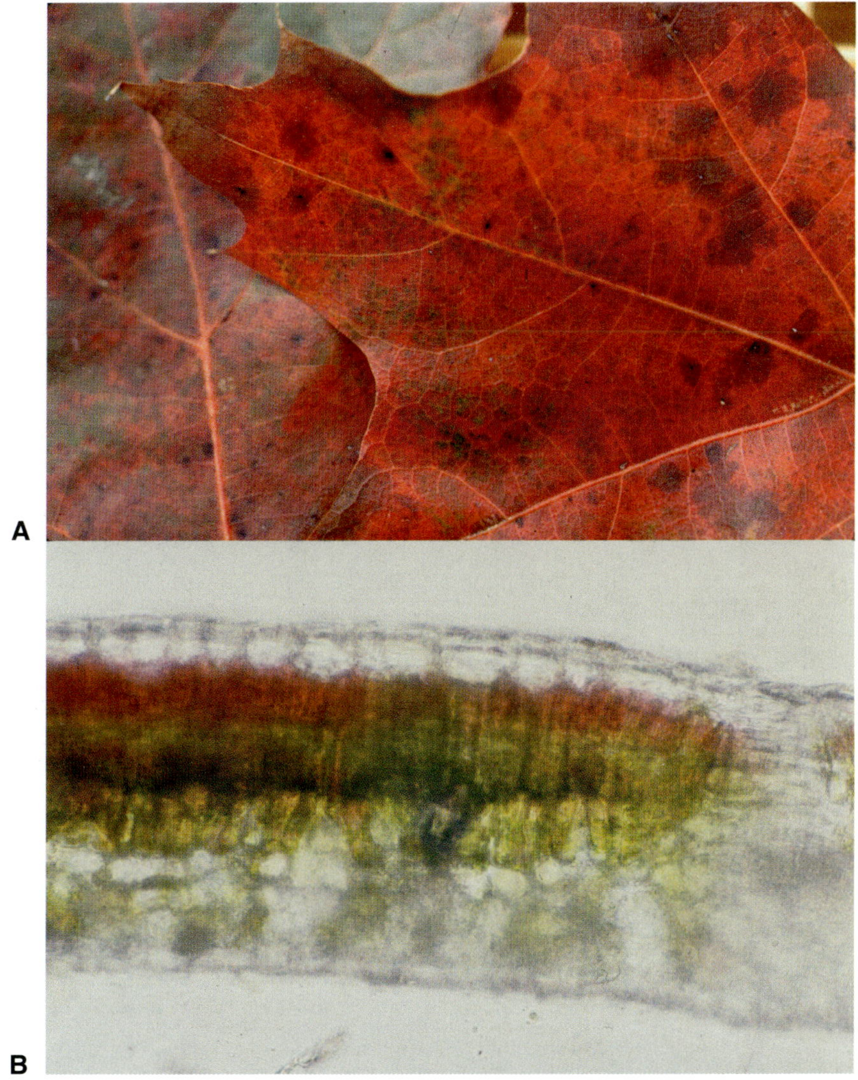

Plate 7. Leaf senescence in the red oak, *Quercus rubra*, photographed at the Harvard Forest, in central Massachusetts. A. Whole leaves in senescence. B. Transverse section of senescent leaf, approximately 250 μm thick, revealing the high concentration of anthocyanins in the palisade parenchyma and some residual chlorophyll.

Plate 8. Variation in anthocyanic distribution among leaves of native New Zealand plants. A, Two branches from one tree of *Quintinia serrata*; B, *Elatostema rugosum*; C, *Pseudowintera colorata*; D, *Dacrydium cupressinum*; E, Adaxial (left) and abaxial (right) surfaces of *Aristotelia serrata*; F, Portion of leaf from *Phormium tenax*.

Less dramatic but nonetheless effective examples of drought-induced anthocyanin accumulation are found in the literature. Increases in anthocyanins and/or associated flavonoids have been noted in drought-stressed *Cotinus* spp. (Oren-Shamir, personal communication), *Cucumis sativus* (Zhi-min *et al.*, 2000), *Vaccinium myrtillus* (Laura Jaakola, personal communication), *Pisum* (Balakumar *et al.*, 1993; Allen *et al.*, 1999), *Pinus taeda* (Heikkenen *et al.*, 1986), *Malus* cv. 'Braeburn' (Mills *et al.* 1994), *Populus* spp. (Wettstein-Westersheim and Minelli, 1962), and *Quercus* spp. (Spyropoulos and Mavormmatis, 1978). These latter authors studied *Quercus* native to environments with differing levels of water availability: *Q. ilex* is found in wet evergreen sclerophyllous forests; *Q. robur* lives in typical mesophytic environments, while *Q. coccifera* is indigenous to the hottest, driest ecosystems. While all three species increased foliar anthocyanins as relative water content decreased, *Q. coccifera* accumulated the most anthocyanins and lost the least amount of water.

Plants from a variety of environments appear to utilize foliar anthocyanins to reduce evapotranspiration. Deciduous tropical trees of the genus *Brachystegia* produce spring flushing leaves rich in anthocyanins two months prior to the rainy season; these leaves have been shown to have lower stomatal conductances than the rainy season green leaves (Tuohy and Choinski Jr, 1990; Choinski and Johnson, 1993). Anthocyanin-containing evergreen leaves of *Mahonia*, *Viburnum* and *Rhododendron* generally had a lower water potential than their green counterparts (Kaku *et al.*, 1992). On the other hand, seedlings of *Corylus avellana* known to be deficient in foliar anthocyanins scorch in field conditions and die (Mehlenbacher and Thompson, 1991), perhaps from excessive water loss.

Research in Doug Schemske's lab may provide linkages between floral anthocyanins and soil moisture content (Schemske and Bierzychudek, 2001). The blue-flowered morph of *Linanthus parryae* is more reproductively competitive on drier soils than the white-flowered form. Similarly, *L. parviflorus* produces both pink- and white-flowered morphs; the first is found in serpentine soils, and the second is found in sandstone deposits. Serpentine soils dry out faster than sandstone soils, so again this difference in anthocyanin content may be associated with drought tolerance. (Although these are floral, rather than foliar, anthocyanins, the phenomenon has not been correlated to pollinator attraction and therefore could be functionally similar to leaf osmotic regulators.)

This relationship between anthocyanin content and drought tolerance has been exploited, perhaps inadvertently, in the use of ornamental landscape plants. Red leaves of *Viburnum opulus* have significantly greater osmotic pressure, and presumably less evapotranspiration, than green leaves (Chalker-Scott, unpublished results). Ornamental shrubs with high levels of anthocyanins, such as *Cotinus* and *Photinia*, have been reported to be more tolerant to drought conditions (Diamantoglou *et al.*, 1989; Knox, 1989; Beeson, 1992; Paine *et al.*, 1992). *Photinia*, a genus that often exhibits juvenile reddening, was found to use water more efficiently when compared to other common ornamental species such as *Ilex*, *Rhododendron*, *Pyracantha*, and *Juniperus* (Knox, 1989). Similarly, Paine *et al.* (1992) found that *Photinia* is broadly tolerant to water regimes. Other red or purple cultivars, like *Capsicum annuum* 'Pretty Purple', are known to be more tolerant to water stress than green cultivars (Bahler *et al.*, 1991).

c. Linkages to Ozone-induced Drought. Other conditions besides limited soil water can induce dehydration stress in plant leaf tissues. Ozone, a photochemical air pollutant, is a strong oxidizing agent with negative impacts on leaf integrity. A variety of gymnosperms and angiosperms, including *Cryptomeria japonica* (Sakata, 1996), *Picea abies* (Maier-Maercker, 1997), *Pinus elliottii* var. *elliottii* (Johnson *et al.*, 1995), *Betula pendula* (Gunthardt-Goerg *et al.*, 1993), and *Populus X euramericana* (Landolt *et al.*, 1994) displayed foliar symptoms such as non-functional guard cells, increased leaf conductance, and cellular collapse as a result of ozone exposure. Not surprisingly, these tissues displayed secondary drought stress symptoms as leaf integrity failed. Other research has reported increased foliar anthocyanin content in *Vaccinium myrtillus* (Nygaard, 1994), *Ipomoea purpurea* (Nouchi and Odaira, 1973), and *Zelkova* (Nouchi and Odaira, 1973) in response to ozone exposure. Unfortunately, these latter studies did not determine the impact of increased foliar anthocyanin on leaf water content and subsequent drought tolerance. Obviously this is an area in need of further study.

B. OSMOTIC STRESS

Unlike drought stress, osmotic stress can occur even when environmental water availability is normal. Osmotic stress occurs when soil water contains enough dissolved solutes that transpiration within the plant is impaired. Much of the available information in the literature is found in tissue culture experiments where growth media is manipulated with an osmoticum. Another body of literature investigates halophytes, which are able to tolerate extreme saline conditions. Still other information is found in mineral nutrition studies where specific elements are found at either deficient or toxic levels. All of these experimental conditions have the commonality of reducing the plant's ability to extract water from its external environment and transport it through its tissues.

1. Impact of Osmotic Stress upon Plant Water Relations

Obviously, increased external osmoticum will induce a dehydration strain on plant cells. High solute levels in the external media will pull water from plant tissues and reduce overall water content. This process can be induced in laboratory studies and observed in the field under conditions of high solute concentrations. Resistance to osmotic stress in either circumstance is generally seen in plants that respond by increasing their internal osmoticum, thereby decreasing overall water potential to the point where water loss is minimized. Foliar anthocyanin concentration will *de facto* decrease the osmotic potential (i.e. make it more negative), decreasing the leaf water potential and retaining cellular water.

2. Evidence of Anthocyanin-mediated Osmotic Resistance

a. Sugar and Salt Stress. Cell cultures are ideal experimental units for understanding the direct impact of solutes, such as sugars, on cellular anthocyanin induction. Studies with cell cultures of various species find anthocyanin accumulation resulting from osmotic stress induced by glucose (Tholakalabavi *et al.*, 1994, 1997), sucrose (Cormier *et al.*, 1989; Do and Cormier, 1991a, b; Rajendran *et al.*, 1992; Sakamoto *et al.*, 1994; Decendit and Mérillon, 1996), and mannitol (Do and Cormier, 1991a; Rajendran *et al.*, 1992; Tholakalabavi *et al.*, 1994, 1997; Suzuki, 1995). Presumably, the accumulation of anthocyanins decreases osmotic potential and thereby prevents the loss of turgor.

Experiments in whole-plant systems have shown similar results. *Arabidopsis* (Mita *et al.*, 1997), *Gerbera* flowers (Amariutei *et al.*, 1995), *Hedera helix* leaves (Murray and Hackett, 1991; Murray *et al.*, 1994), *Terminalia catappa* leaves (Dubé *et al.*, 1993), and several other species (as reviewed by Weiss, 2000) all accumulated anthocyanins when grown in the presence of various sugars. Sugar-induced accumulation of anthocyanins is indicated as well in work by Jeannette *et al.* (2000). In this study, heat-girdled *Zea mays* leaves accumulated anthocyanins and remained turgid. Since heat girdling was found to impair translocation in phloem, but not xylem transport, the expected increase of foliar sugars could certainly induce the synthesis of anthocyanins.

Exposure to salts also induces anthocyanin accumulation in roots (Kaliamoorthy and Rao, 1994), stems (Dutt *et al.*, 1991), and leaves (Ramanjulu *et al.*, 1993) of various species. Kennedy and DeFilippis (1999), in comparing the salt tolerant, anthocyanic *Grevillea ilicifolia* to the salt sensitive *G. arenaria*, admit thinking that anthocyanins are 'not essential for growth and survival of plants', but conclude they are acting as osmoregulators.

Of more ecological interest are the halophytes which are able to tolerate extremely saline conditions on land and in water. Eight mangrove species, native to marine habitats, were found to have high anthocyanins in their evergreen foliage during the months of October and November (Oswin *et al.*, 1994). Given that these are the months preceding the monsoon season, it is likely that the salinity is quite high and therefore so are foliar anthocyanins. Terrestrial halophytes include *Mesembryanthemum crystallinum*, also known as the ice plant. These and other members of the Caryophyllales contain betalains instead of anthocyanins (Vogt *et al.* 1999). These closely related pigments apparently have the same ecophysiological function in water retention as the anthocyanins. When subjected to drought or water stress, *M. crystallinum* increases the size of bladder cells at the leaf tips (Cockburn *et al.*, 1996), which then accumulate betalains (Vogt *et al.*, 1999).

b. Nutrient Stress. Nutrient deficiencies have long been associated with anthocyanin accumulation. The best known example is foliar reddening from phosphorus deficiency, identified in early nutrient exclusion experiments (McMurtrey, 1938). Increases in foliar anthocyanins have been reported in phosphorus-deficient *Acer saccharum* (Bernier and Brazeau, 1988), *Arabidopsis* (Zakhleniuk *et al.*, 2001), *Cunninghamia lanceolata* (Kao *et al.*, 1973), *Daucus carota* (Rajendran *et al.* 1992), *Eucalyptus grandis* (Lacey *et al.*, 1966), *Ixora* (Broschat, 2000), and *Nepenthes* (Moran and Moran, 1998). An assortment of other nutritional deficiencies have also been associated with foliar anthocyanin accumulation, including boron deficiency in *Eucalyptus globulus* (Dell and Malajczuk, 1994), magnesium deficiency in *Trifolium subterraneum* (Michalk and Huang, 1992), nitrogen deficiency in *Daucus carota* (Rajendran *et al.* 1992), *Eucalyptus nitens* (Close *et al.*, 2001), and *Vitis vinifera* (Do and Cormier, 1991b), sulfur deficiency in *Dendranthema grandiflora* (Huang *et al.*, 1997), and zinc deficiency in *Eucalyptus marginata* (Wallace *et al.*, 1986).

These strong correlations between nutritional stress and anthocyanin accumulation can be linked to water relations and morphological leaf characteristics. Early studies (Fritzsche and Kemmer, 1959; Kemmer and Fritzsche, 1961) found positive correlations between nutrient deficiency, petiole hairiness, and anthocyanin content in *Populus* spp.; these leaf characteristics have been previously associated with drought tolerance. Mineral stressed soils were seen to induce cuticular thickening, anthocyanin accumulation, and premature fall senescence in *Quercus palustris* (Boyer, 1988). Phosphorus deficiency induced water stress in *Morus alba* (Sharma, 1995) and inhibited root conductivity in *Fraxinus pennsylvanica* (Andersen *et al.*, 1989); presumably the latter effect would induce water stress as well. Feller (1996) goes as far as to suggest that the sclerophyllous leaves of *Rhizophora mangle* are actually an adaptation to phosphorus deficient soils rather than a response to drought.

The causal relationship behind nutritional deficiencies, anthocyanin accumulation, and water relations has not been sufficiently explored. It is likely that many of these deficiencies are either directly or indirectly associated with water uptake or water retention, leading to the development of drought-associated characteristics such as pubescence, cuticular thickening, or anthocyanin accumulation.

c. *Metal Toxicity.* At the other end of the nutritional spectrum are stresses associated with mineral toxicities. Often, these might be essential elements in greater than needed quantities, but non-essential minerals, especially heavy metals, can also have negative effects. Anthocyanins are known to bind many heavy metals including magnesium and aluminum, which may help plants tolerate serpentine soils and other soils containing high metal concentrations. For example, the pink-flowered form of *Linanthus parviflorus* is tolerant of serpentine soils while white-flowered populations are not (Schemske and Bierzychudek, 2001). Although these are floral and not foliar anthocyanins, the function of anthocyanins in these tissues appears to be related to heavy metal tolerance rather than pollinator attraction.

Similar anthocyanin-mediated metal tolerance appears in the published literature. In an early study, lead chloride-treated seedlings of *Acer rubrum* accumulated anthocyanins in their leaves (Davis and Barnes, 1973). Other experimental efforts have linked anthocyanin accumulation to exposure to heavy metals, including aluminum (Escobar-Munera, 1988), molybdenum (Hale *et al.*, 2001), and calcium, cobalt, iron, manganese, vanadium, and zinc (Suvarnalatha *et al.*, 1994). Conversely, interference with ethylene biosynthesis and subsequent phenolic production was thought to be the reason for the inhibition of anthocyanin accumulation by cobalt in *Terminalia catappa* leaf disks (Dubé *et al.*, 1993).

The mechanism(s) behind heavy metal tolerance is not clearly understood. A possible phytochelating effect is described by Hale *et al.* (2001), whereby anthocyanins bind to molybdenum and are sequestered in the vacuoles of peripheral tissues. An inability to form phytochelators (including anthocyanins) was found in *Arabidopsis* mutants with hypersensitivity to cadmium exposure (Xiang *et al.*, 2001). Other evidence suggests that heavy metals affect plant water relations. Aluminum exposure is known to cause drought stress in *Glycine max* (Foy *et al.*, 1993; Spehar and Galwey, 1996). A drought-tolerant cultivar of *Triticum durum* was less affected by nickel exposure than a drought-sensitive cultivar (Pandolfini *et al.*, 1996). Nickel exposure causes decreased water potential and relative water content, so drought resistance is thought to enhance nickel resistance. Given that one of these metals (aluminum) has also been linked to increased anthocyanin production (Escobar-Munera,

1988), it is likely that osmotic regulation by anthocyanins plays a role in tolerating heavy metal exposure.

C. COLD STRESS

At first glance, cold stress appears to have little in common with either drought or osmotic stress. When plant tissues are subjected to rapid decreases in temperature, the formation of intracellular ice expands and lyses cell membranes. Such conditions, however, do not often occur in nature. Most damage caused to plant tissues by cold temperatures in nature is instead caused by freeze-induced dehydration, not by physical damage from ice crystals. Therefore, the stress induced on the plant is again related to water stress, albeit indirectly.

1. Impact of Cold Temperatures upon Plant Water Relations

Under normal rates of temperature depression, extracellular water freezes first as it is more dilute and water molecules can easily organize into the crystalline structure necessary for ice formation. This creates a water deficit in the extracellular spaces, which in turn pulls water from adjacent cells. The more dilute the cell sap, the more water can be pulled away, resulting in intracellular dehydration stress and eventual plasmolysis. Therefore, it would be advantageous for leaf tissues exposed to frosts or freezes to maintain a lower osmotic potential via increased solute accumulation.

2. Evidence of Anthocyanin-mediated Freezing Resistance

In an earlier review (Chalker-Scott, 1999), the possible relationship between cold hardiness and anthocyanins has been described in greater detail. Briefly, low temperature has been shown to induce anthocyanin synthesis and accumulation in the leaves of a number of species, including *Calluna vulgaris* (Foot *et al.*, 1996), *Centaurea cyanus* (Kakegawa *et al.*, 1987), *Cornus stolonifera* (Van Huystee *et al.*, 1967), *Cotinus* (Oren-Shamir and Levi-Nissim, 1997a, b), *Malus domestica* (Leng *et al.*, 2000), *Pinus* spp. (Camm *et al.*, 1993; Krol *et al.*, 1995), *Poncirus trifoliata* (Tignor *et al.*, 1997), and *Prunus persica* (Leng *et al.*, 2000). Other studies have associated more general autumnal or winter conditions with the foliar anthocyanins of *Acer* spp. (Ishikura, 1973; Ji *et al.*, 1992), *Euonymous* spp. (Ishikura, 1973), *Hedera helix* (Parker, 1962; Murray *et al.*, 1994), *Mahonia repens* (Grace *et al.*, 1998), *Pinus banksiana* (Nozzolillo *et al.*, 1990), *Rhus* spp. (Ishikura, 1973), and *Terminalia catappa* (Dubé *et al.*, 1993).

Comparison of populations from distinct provenances has revealed differential anthocyanin accumulation. Northernmost populations of

Carya illinoinensis (Wood *et al.*, 1998) and *Populus trichocarpa* (Howe *et al.*, 1995) contained greater foliar anthocyanins than more southern populations. Likewise, *Pinus contorta* from different localities in Norway accumulate more anthocyanins as altitude and latitude increase (Dietrichson, 1970). Furthermore, red- and purple-leaved varieties of *Acer palmatum* native to northern regions lose their pigmentation when grown in southern climes (Deal *et al.*, 1990). Given the generally colder and shorter growing season in higher altitude or latitude environments, it is likely that the relationship between anthocyanins and cold hardiness is more than coincidental.

Do foliar anthocyanins actually increase cold hardiness? Much of the evidence in the literature suggests that they do, but few researchers have asked this specific question. Decades ago, Parker (1962) linked anthocyanin appearance and disappearance to cold hardiness in *Hedera helix* leaves. Es'kin (1960) postulated a protective function of foliar anthocyanins in avoiding frost damage. Grace *et al.* (1998) noted seasonally induced anthocyanin accumulation and disappearance in *Mahonia repens*. Recently, McKown *et al.* (1996) described four *Arabidopsis* mutants deficient in freezing tolerance were unable to accumulate anthocyanins, suggesting some commonality between anthocyanin biosynthesis and freezing tolerance. Finally, the seasonally-induced appearance and disappearance of anthocyanins in *Pinus banksiana* seedlings led Nozzolillo *et al.* (1990) to suggest that anthocyanin accumulation in this species was directly associated with exposure to low temperature rather than any other environmental factor.

3. *Anthocyanic Osmoregulation and Avoidance of Freeze Damage*

Young tissues are particularly sensitive to frost damage and therefore require protection from unseasonably cold temperatures, especially those associated with spring frosts. Expanding leaves, with their high water content and lack of cuticular protection, can be damaged or killed by ice formation on the leaf surface and subsequent nucleation events within tissues. The presence of increased osmoticum within tissues would allow leaves to supercool and avoid internal freezing events. This phenomenon was noted in expanding leaves of *Eucalyptus urnigera* (Thomas and Barber, 1974); likewise, anthocyanins were found to accumulate in the young leaves of *Zea mays* following cold spells (Crookston, 1983).

There is little in the literature that associates water relations, anthocyanins, and cold temperatures, however. Anthocyanin-containing leaves of winter-hardy *Mahonia*, *Rhododendron*, and *Viburnum* were shown to have a lower water potential during the winter than green leaves of the same species (Kaku *et al.*, 1992). Similarly, Tignor *et al.* (1997) found a depression of water potential concomitant with an increase in anthocyanin content and cold hardiness development in *Poncirus trifoliata* following low temperature exposure. Further studies are obviously necessary.

In an earlier review (Chalker-Scott, 1999), I hypothesized that low temperature exposure would induce a small but significant increase in hardiness via osmotic regulation: more solutes (e.g. anthocyanins) in the vacuole depress the freezing point. This small increase in hardiness would be enough to protect young tissues from frost damage in late spring. In particular, the accumulation of anthocyanins in epidermal vacuoles would prevent their freezing, especially from leaf surface nucleators. This phenomenon would also protect deciduous leaves from early autumn frosts – a physiologically important time during which substances are mobilized for winter storage. Perennial tissues then show a second, more significant increase in cold hardiness (seen during the winter) several weeks after exposure, which may or may not be related to anthocyanins.

D. ANOXIC STRESS

Anoxia and hypoxia are defined as an absence or deficiency of oxygen, respectively. Anoxic or hypoxic soils are commonly found in wetland environments, where water fills available pore spaces within the soil at the expense of oxygen and other gases. Obligate and facultative wetland plants are adapted to these conditions and have a variety of morphological and biochemical modifications to avoid root anoxia. Conversely, plants that are not normally found in anoxic soils are often unable to adapt to low oxygen levels and roots become less functional. As a result, the above ground portions of the plant are subjected to an indirect water stress.

1. Impact of Anoxia upon Plant Water Relations

Anoxic soil conditions can occur as a function of high clay content, prolonged flooding, or compaction. Though root water uptake decreases, foliar transpiration continues and subsequently the above-ground portions of the plant will become dehydrated. Thus, any mechanism that helps maintain leaf water under these conditions will increase survival.

2. Evidence of Anthocyanin-mediated Anoxia Resistance

Anthocyanins will accumulate in plants subjected to root anoxia. Young *Arbutus menziesii* grown in compacted, waterlogged clay soils developed more foliar anthocyanins than those grown in well-drained, sandy loam (Cahill and Chalker-Scott, 2000). Similarly, *Acer rubrum* seedlings grown on silt loam accumulated more anthocyanins than those grown on sandy loam (Davis and Barnes, 1973). Flooded *Malus* and *Pyrus* trees showed anthocyanin increases in their leaves (Anderson *et al.*, 1984), probably in response to the secondary drought stress imposed upon leaves by depressed root function. Moreover, *Pyrus* subjected to flooding was able to avoid abscission over a period of several months (Anderson *et al.*, 1984), perhaps because the foliar anthocyanins depressed osmotic potential and reduced leaf evaporation.

III. RELATIONSHIPS AMONG ANTHOCYANINS, DEVELOPMENTAL STATE, AND WATER RELATIONS

As reviewed to this point, there is a great deal of information on anthocyanin induction in leaf tissues in a variety of plant species under a variety of environmental conditions. Many of these environmental conditions can be linked to plant water relations. There are instances, however, of anthocyanin accumulation where water stress appears to be less important: juvenile leaf reddening is the principal example. In other cases, anthocyanin accumulation is associated with autumnal or winter foliage during times of the year when water uptake and transpiration are limited. Are there relationships between leaves at these different developmental stages and anthocyanin and water content?

A. JUVENILE REDDENING, LEAF EXPANSION AND OSMOREGULATION

1. Occurrence of the Phenomenon

One of the most striking phenomena associated with anthocyanins is the juvenile reddening exhibited by a variety of plant species. Young expanding leaves accumulate anthocyanins transiently, losing these pigments once they have reached full size. Juvenile reddening is particularly noticeable in tropical evergreens and has been reported in several studies (Lee and Lowry, 1980; Tuohy and Choinski, 1990; Whatley, 1992; Nii *et al.*, 1995; Woodall *et al.*, 1998). Representatives are found in both emergent as well as understory species and include agriculturally and horticulturally important genera such as *Ficus*, *Coffea*, *Durio*, and *Mangifera*. Opler *et al.* (1980) estimated that between 20–40% of the woody species in a typical humid, tropical forest contain juvenile anthocyanins. More recently, Lee and Collins (2001) identified 179 tropical angiosperm species that exhibit juvenile reddening.

A number of plants native to other environments also exhibit juvenile anthocyanins, though the occurrence appears to be less common than that of the tropical rainforest. The temperate Aceraceae are common juvenile accumulators (Ji *et al.*, 1992) as are several *Eucalyptus* species (Hillis, 1955; Thomas and Barber, 1974). Other temperate species exhibiting juvenile reddening include *Liquidambar styraciflua* (Dillenberg *et al.*,

1995), *Quercus marilandica* (Choinski and Wise, 1999), and *Photinia* (Knox, 1989; Paine *et al.*, 1992; Plate 4). Finally, *Brachystegia*, a deciduous tree genus native to tropical savannas, accumulates anthocyanins in its young leaves (Tuohy and Choinski Jr, 1990; Choinski and Johnson, 1993).

2. Functional Roles of Juvenile Anthocyanins

The role of anthocyanins in expanding leaf and floral tissues has not been well defined. While floral anthocyanins have been postulated to act as pollinator attractants, expanding floral tissues are not able to utilize pollinating vectors. Therefore, these pigments may play a different role during tissue expansion. In both young leaves and flowers it is crucial to maintain high turgor to ensure optimal expansion. Turgor pressure is the force behind cell wall expansion, and if a suitable gradient is not established young tissues may not achieve full expansion. Bieleski (1993) suggested that increased osmoticum in *Hemerocallis* petals was the driving force behind petal expansion. Weiss (2000) noted that flower buds are a major sink for sugar and that flowering could be delayed if photo-assimilates were not available. Anthocyanins might facilitate the transport of sugars to expanding flowers since they exist almost exclusively as glycosides. This same process would occur during leaf expansion.

This latter hypothesis would explain why so many understory rainforest species exhibit juvenile reddening. While some authors have postulated a primarily photoprotective function for anthocyanins (Hoch *et al.*, 2001), this hypothesis does not hold up under the reduced light environment of tropical forests. Relative humidity is high under the canopy and water movement through subcanopy plants can be reduced. If high osmotic conditions are established in juvenile tissues – such as the accumulation of water-soluble pigments like anthocyanins and betacyanins – then an appropriate gradient can be established to ensure increased water uptake, high turgor, and cell wall expansion.

Tropical species in drier environments, such as *Brachystegia* (Tuohy and Choinski Jr, 1990; Choinski and Johnson, 1993), may also rely upon anthocyanic osmoregulation. Given the lack of cuticle or other epidermal protection, expanding leaf tissues are more likely than mature leaves to lose water. *Brachystegia* produce anthocyanin-rich leaves during the months prior to the rainy season; after the rains begin, the leaves turn green. The stomatal conductance of the red, pre-rain leaves has been found to be lower than that of green leaves (Tuohy and Choinski Jr, 1990; Choinski and Johnson, 1993). The ability of anthocyanins to decrease leaf osmotic potential would allow plants to retain water more efficiently than leaves without lowered potential.

This same ability to retain water during leaf expansion would benefit temperate zone plants. Many temperate zone species, such as *Hedera helix* (Murray and Hackett, 1991), *Photinia* (Knox, 1989; Paine *et al.*, 1992; Plate 4), and *Quercus marilandica* (Choinski and Wise, 1999) exhibit juvenile reddening and, in the case of *Q. marilandica*, also have reduced transpiration rates and stomatal conductance (Choinski and Wise, 1999). This last bit of evidence also suggests an osmoregulatory ability of anthocyanins in these young leaves.

Temperate zone plants not only need to maintain high turgor for leaf expansion, but need to protect tender growing tissues from late season freezes. Unlike mature tissues, expanding leaves cannot cold-harden by lignifying their cell walls. Anthocyanin accumulation by epidermal cells in these latter tissues would decrease the osmotic potential of the epidermis and delay freezing via surface nucleators, thus protecting the leaves from late spring frosts. Thomas and Barber (1974) reported supporting evidence from research on leaves of *Eucalyptus urnigera*. Rapid temperature drops injured expanding *E. urnigera* leaves, but if temperature rates fell at more natural rates expanding leaves survived. Such supercooling ability could be very likely due to the anthocyanin content of the young leaves of this species and would be an obvious benefit to temperate zone plants that produce leaves when the danger of frost is still present.

B. AUTUMNAL COLORATION, LEAF SENESCENCE, AND FROST PROTECTION

> … the accumulation of anthocyanins in the vacuoles of leaf cells seems to represent a kind of extravagancy without a vital function … Whereas they seem to have no function in trees, they contribute to the autumnal feast for the human eyes.
>
> (Matile, 2000)

While the above sentiment would be understandable (if erroneous) a hundred years ago, it is surprising to read it in a current publication. The accumulation of anthocyanins in temperate, senescing deciduous leaves has been intensively studied both descriptively and experimentally. It is known that anthocyanins are synthesized *de novo* in response to the shorter days and cooler nights of autumn. The ability of anthocyanins to bind with otherwise reactive sugars such as glucose makes these pigments logical candidates for carbohydrate transport during senescence. Indeed, the presence of foliar sugars has been shown to induce anthocyanin synthesis as discussed earlier. In some studies, such as one performed with leaf disks of *Terminalia catappa*, this induction has been tied to pigment accumulation during autumn senescence (Dubé *et al.*, 1993). Many other species, including members of the Aceraceae (Ishikura, 1973; Ji *et al.*, 1992), *Cornus stolonifera* (Feild *et al.*, 2001), *Euonymus* and *Rhus* (Ishikura, 1973) also show anthocyanin accumulation in senescing leaf tissue.

The phenomenon of autumnal reddening has been attributed to anthocyanic photoprotection of senescing leaves (Feild *et al.*, 2001; Hoch *et al.*, 2001); there is, however, an alternate hypothesis that merits exploration. It is apparent that most, if not all, leaf tissues with autumnal anthocyanins sequester the pigments in their upper epidermal or mesophyll layers (Hoch *et al.*, 2001). Additionally, anthocyanin accumulation may only occur in outer canopy leaves (Feild *et al.*, 2001). The adaxial leaf surfaces, especially those in the outer canopy, are exposed to warm, high light conditions during the day and as their cuticles erode would be more likely to lose water through evapotranspiration. There is, in fact, a gradual water loss associated with leaf senescence. Anthocyanins in these adaxial tissues would depress the osmotic potential of the leaf so as to decrease water loss and support those biochemical pathways associated with programmed senescence. Furthermore, foliar anthocyanins located in upper epidermal or mesophyll layers would serve to depress the freezing point of water in these tissues, reducing the likelihood of the internal spread of ice from nucleating events on outer canopy leaf surfaces during frosty autumn nights.

C. WINTER ANTHOCYANINS, PERENNIAL TISSUES, AND COLD HARDINESS

In addition to autumnal frosts, temperate zone perennials undergo an annual low temperature stress which can manifest itself as freeze-induced dehydration. Above-ground tissues are especially at risk, and none more so than evergreen leaves. While dormant structures, such as buds, have various mechanisms by which to resist dehydration stress, leaves remain functional, hydrated, and susceptible to freeze damage. It is not surprising that many of these evergreen perennials are found to contain foliar anthocyanins during the winter months. Many of these species lose their red coloration in the warmer spring and summer months (Parker, 1962; Nozzolillo *et al.*, 1990; Murray *et al.*, 1994; Grace *et al.*, 1998; Leng *et al.*, 2000), providing more evidence of the importance of these pigments during freezing conditions.

As discussed earlier, foliar anthocyanins located in upper epidermal or mesophyll layers would serve to depress the freezing point of water in these tissues, reducing the likelihood of the internal spread of ice from nucleating events on the leaf surface. Additionally, the presence of cellular anthocyanins would decrease the solute potential and inhibit water migration into extracellular spaces. As the threat of spring frost declines, anthocyanins are catabolized and leaves return to their normal green condition (assuming that water conditions remain optimal in leaf tissues).

IV. ANTHOCYANIC TRANSIENCE AND VARIABILITY IN PLANT WATER STATUS

Anthocyanins often appear transiently at specific developmental stages and may be induced by a number of environmental factors, including visible and UV-B radiation, cold temperatures, drought, salinity, mineral toxicity, or nutrient stress (Chalker-Scott, 1999). Given their potential multifunctionality, the production and localization of anthocyanins in leaf tissues may help seedlings (Krol *et al.*, 1995) or expanding tissues establish under a suite of sub-optimal environmental conditions, many of which will directly or indirectly influence plant water relations. These possible functions are summarized below.

A. REDUCTION OF EVAPOTRANSPIRATION

Pubescence, thickened cuticles and cell walls all help decrease foliar water loss in mature tissues. Unlike these morphological modifications, however, the accumulation of anthocyanins can be transient and is therefore a more flexible adaptation. Furthermore, it is more likely that developing leaves, which lack these morphological modifications, would rely on vacuolar substances to modify water relations.

B. OSMOTIC REGULATION

Anthocyanins are highly water soluble, especially as glycosides, and are usually found in vacuoles. The increased accumulation of solute in any plant cell will allow it to maintain tolerable water conditions even under high levels of external osmoticum. This has particularly important implications for plants resistant to saline habitats.

C. ANTIOXIDANT BEHAVIOR

Plants in sub-optimal water conditions not only experience dehydration stress but increased risk of oxidative stress from free radicals. Most, if not all, environmental stresses occur with a concomitant production of oxygen radicals or other reactive substances that can cause membrane damage unless attenuated by antioxidants. Given the antioxidative properties of anthocyanins (discussed by Gould *et al.* in this volume), their presence could protect sensitive structures such as membranes (Leng *et al.*, 2000) or chlorophyll from degradation and also increase the percentage of bound water within the cells. This phenomenon in the extreme has been documented in resurrection plants.

D. SOLAR PROTECTION

Although solar protection may not be a universal feature in all anthocyanic leaf tissue, this attribute nonetheless would serve to diminish the possibility of photooxidative stress, especially during high light, low temperature periods. The role of anthocyanins in solar protection is discussed in the chapter by Timmins and colleagues, in this volume.

E. SUGAR TRANSPORT

Anthocyanins are generally glycosylated and therefore extremely soluble compounds (Robinson, 1991). This capability allows anthocyanins to bind and transport reactive monosaccharides produced during developmentally or environmentally critical stages. In addition to their presence in senescing leaves and expanding leaves and flowers, anthocyanins in ray parenchyma of cold hardy trees (Schmucker, 1947) might very well serve in this capacity.

F. HEAVY METAL CHELATION

Since anthocyanins can apparently bind and sequester heavy metals, many of which cause secondary drought stress, their presence may allow plants to survive in metal-rich soils.

G. FREEZE AVOIDANCE

When the water potential of adaxial leaf surfaces is lowered, two environmental stresses can be avoided: ice nucleation via freezing events on the leaf surface and freeze-induced drought. During winter freezes this could be particularly important in protecting woody evergreens from freeze-induced dehydration of sensitive parenchyma cells both in the mesophyll and in xylem rays.

A cautionary note needs to be inserted lest one regards anthocyanins to be unique in their transitory protective nature. Because these pigments are visible to the human eye, they have attracted more attention than their numerous relatives in the phenolic acid family. It may well be that less visible, and hence less studied, chemical relatives show the same type of transience in plant tissues relative to developmental or environmental cues.

REFERENCES

Allen, D. J., Nogués, S., Morison, J. I. L., Greenslade, P. D., McLeod, A. R. and Baker, N. R. (1999). A thirty percent increase in UV-B has no impact on photosynthesis in well-watered and droughted pea plants in the field. *Global Change Biology* **5**, 235–244.

Amariutei, A., Alexe, C. and Burzo, I. (1995). Physiological and biochemical changes of cut *Gerbera* inflorescences during vase life. *Acta Horticulturae* **405**, 372–380.

Andersen, C. P., Sucoff, E. I., Dixon, R. K. and Markhart, A. H. III (1989). Effects of phosphorus deficiency on root hydraulic conductivity in *Fraxinus pennsylvanica*. *Canadian Journal of Botany* **67**, 472–476.

Anderson, P. C., Lombard, P. B. and Westwood, M. N. (1984). Leaf conductance, growth, and survival of willow and deciduous fruit tree species under flooded soil conditions. *Journal of the American Society for Horticultural Science* **109**, 132–138.

Bahler, B. D., Steffen, K. L. and Orzolek, M. D. (1991). Morphological and biochemical comparison of a purple-leafed and a green-leafed pepper cultivar. *HortScience* **26**, 736.

Balakumar, T., Hani Babu Vincent, V. and Paliwal, K. (1993). On the interaction of UV-B radiation (280–315 nm) with water stress in crop plants. *Physiologia Plantarum* **87**, 217–222.

Beeson, R. C. (1992). Restricting overhead irrigation to dawn limits growth in container-grown woody ornamentals. *HortScience* **27**, 996–999.

Bernier, B. and Brazeau, M. (1988). Nutrient deficiency symptoms associated with sugar maple dieback and decline in the Quebec Appalachians. *Canadian Journal of Forest Research* **18**, 762–767.

Bieleski, R. L. (1993). Fructan hydrolysis drives petal expansion in the ephemeral daylily flower. *Plant Physiology* **103**, 213–219.

Boyer, M., Miller, J., Belanger, M. and Hare, E. (1988). Senescence and spectral reflectance in leaves of Northern Pin Oak (*Quercus palustris* Muenchh.) *Remote Sensing of Environment* **25**, 71–87.

Broschat, T. K. (2000). Potassium and phosphorus deficiency symptoms of ixora. *HortTechnology* **10**, 314–317.

Burger, J. and Edwards, G. E. (1996). Photosynthetic efficiency, and photodamage by UV and visible radiation, in red versus green leaf coleus varieties. *Plant Cell Physiology* **37**, 395–399.

Cahill, A. and Chalker-Scott, L. (2000). The role of soil environment in *Arbutus menziesii* (Pacific madrone) seedling success. *American Nurseryman* **193**, 26–34.

Camm, E. L., McCallum, J., Leaf, E. and Koupai-Abyazani, M. R. (1993). Cold-induced purpling of *Pinus contorta* seedlings depends on previous daylength treatment. *Plant, Cell and Environment* **16**, 761–764.

Chalker-Scott, L. (1999) Environmental significance of anthocyanins in plant stress responses. *Photochemistry and Photobiology* **70**, 1–9.

Choinski, J. S. and Johnson, J. M. (1993). Changes in photosynthesis and water status of developing leaves of *Brachystegia spiciformis* Benth. *Tree Physiology* **13**, 17–27.

Choinski, J. S. and Wise, R. R. (1999). Leaf growth and development in relation to gas exchange in *Quercus marilandica* Muenchh. *Journal of Plant Physiology* **154**, 302–309.

Close, D. C., Davies, N. W. and Beadle, C. L. (2001). Temporal variation of tannins (galloylglucoses), flavonols and anthocyanins in leaves of *Eucalyptus nitens* seedlings: implications for light attenuation and antioxidant activities. *Australian Journal of Plant Physiology* **28**, 269–278.

Cockburn, W., Whitelam, G. C., Broad, A. and Smith, J. (1996). The participation of phytochrome in the signal transduction pathway of salt stress responses in *Mesembryanthemum crystallinum* L. *Journal of Experimental Botany* **47**, 647–653.

Cormier, F., Crevier, H. A. and Do, C. B. (1989). Effect of sucrose concentration on the accumulation of anthocyanins in grape (*Vitis vinifera*) cell suspension. *Canadian Journal of Botany* **68**, 1822–1825.

Crookston, R. K. (1983). Purple corn: conspicuous color in spring not always bad. *Crops and Soils Magazine* **35**, 16–18.

Curtis, J. D., Lersten, N. R. and Lewis, G. P. (1996). Leaf anatomy, emphasizing unusual 'concertina' mesophyll cells, of two east African legumes (Caesalpinieae, Caesalpiniodideae, Leguminosae). *Annals of Botany* **78**, 55–59.

Davis, J. B. and Barnes, R. L. (1973). Effects of soil-applied fluoride and lead on growth of loblolly pine and red maple. *Environmental Pollution* **5**, 35–44.

Deal, D. L., Raulston, J. C. and Hinesley, L. E. (1990). Leaf color retention, dark respiration, and growth of red-leafed Japanese maples under high night temperatures. *Journal of the American Society for Horticultural Science* **115**, 135–140.

Decendit, A. and Mérillon, J. M. (1996). Condensed tannin and anthocyanin production in *Vitis vinifera* cell suspension cultures. *Plant Cell Reports* **15**, 762–765.

Dell, B. and Malajczuk, N. (1994). Boron deficiency in eucalypt plantations in China. *Canadian Journal of Forest Research* **24**, 2409–2416.

Diamantoglou, S., Rhizopoulou, S. and Kull, U. (1989). Energy content, storage substances, and construction and maintenance costs of Mediterranean deciduous leaves. *Oecologia* **81**, 528–533.

Dietrichson, J. (1970). Geographic variation in *Pinus contorta*: a study with a view to the use of this species in Norway. *Meddelelser fra det Norske Skogforsøksvesen* 28, 111–140.

Dillenberg, L. R., Sullivan, J. H. and Teramura, A. H. (1995). Leaf expansion and development of photosynthetic capacity and pigments in *Liquidambar styraciflua* (Hamamelidaceae) – effects of UV-B radiation. *American Journal of Botany* **82**, 878–885.

Do, C. B. and Cormier, F. (1991a). Accumulation of peonidin 3-glucoside enhanced by osmotic stress in grape (*Vitis vinifera* L.) cell suspension. *Plant Cell, Tissue, and Organ Culture* **24**, 49–54.

Do, C. B. and Cormier, F. (1991b). Effects of low nitrate and high sugar concentrations on anthocyanin content and composition of grape (*Vitis vinifera* L.) cell suspension. *Plant Cell Reports* **9**, 500–504.

Dubé, A., Bharti, S. and Laloraya, M. M. (1993). Inhibition of anthocyanin synthesis and phenylalanine ammonia-lyase activity by Co^{2+} in leaf disks of *Terminalia catappa*. *Physiologia Plantarum* **88**, 237–242.

Dutt, S. K., Bal, A. R. and Bandyopadhyay, A. K. (1991). Salinity induced chemical changes in *Casuarina equisetifolia* Forst. *Epyptian Journal of Soil Science* **31**, 57–63.

Escobar-Munera, M. L. (1988). Aluminium toxicity in *Eucalyptus grandis* and its possible correction. *Informa Servicio Nacional de Proteccion Forestal* **2**, 49–62.

Es'kin, B. I. (1960). Anthocyanin and plant frost resistance. *Botanical Science* **130**, 58–60.

Farrant, J. M. (2000). A comparison of mechanisms of desiccation tolerance among three angiosperm resurrection plant species. *Plant Ecology* **151**, 29–39.

Feild, T. S., Less, D. W. and Holbrook, N. M. (2001). Why leaves turn red in autumn. The role of anthocyanins in senescing leaves of red-osier dogwood. *Plant Physiology* **127**, 566–574.

Feller, I. C. (1996). Effects of nutrient enrichment on leaf anatomy of dwarf *Rhizophora mangle* L. (red mangrove). *Biotropica* **28**, 13–22.

Foot, J. P., Caporn, S. J. M., Lee, J. A. and Ashenden, T. W. (1996). The effect of long-term ozone fumigation on the growth, physiology and frost sensitivity of *Calluna vulgaris*. *New Phytologist* **133**, 503–511.

Foy, C. D., Carter, T. E. Jr, Duke, J. A. and Devine, T. E. (1993). Correlation of shoot and root growth and its role in selecting for aluminum tolerance in soybean. *Journal of Plant Nutrition* **16**, 305–325.

Fritzsche, K. and Kemmer, C. (1959). The effect of Ca on growth and development of poplar cuttings with special reference to pH. *Wissenschaftliche Abhandlungen / Deutsche Akademie der Landwirtschaftswissenschaften zu Berlin* **40**, 135–169

Grace, S. C., Logan, B. A., Adams, W. W. III (1998). Seasonal differences in foliar content of chlorogenic acid, a phenylpropanoid antioxidant, in *Mahonia repens*. *Plant, Cell and Environment* **21**, 513–521.

Gunthardt-Goerg, M. S., Matyssek, R., Scheidegger, C. and Keller, T. (1993). Differentiation and structural decline in the leaves and bark of birch (*Betula pendula*) under low ozone concentrations. *Trees: Structure and Function* **7**, 104–114.

Hale, K. L., McGrath, S. P., Lombi, E., Stack, S. M., Terry, N., Pickering, I. J., George, G. N. and Pilon-Smits, E. A. H. (2001). Molybdenum sequestration in *Brassica* species. A role for anthocyanins? *Plant Physiology* **126**, 1391–1402.

Heikkenen, H. J., Scheckler, S. E., Egan, P. J. J. Jr and Williams, C. B. Jr (1986). Incomplete abscission of needle clusters and resin release from artificially water-stressed loblolly pine (*Pinus taeda*): a component for plant–animal interactions. *American Journal of Botany* **73**, 1384–1392.

Hillis, W. E. (1955). Formation of leuco-anthocyanins in eucalypt tissues. *Nature* **175**, 597–598.

Hoch, W. A., Zeldin, E. L. and McCown, B. H. (2001). Physiological significance of anthocyanins during autumnal leaf senescence. *Tree Physiology* **21**, 1–8.

Howe, G. T., Hackett, W. P., Furnier, G. R. and Klevorn, R. E. (1995). Photoperiodic responses of a northern and southern ecotype of black cottonwood. *Physiologia Plantarum* **93**, 695–708.

Huang, L. C., Paparozzi, E. T. and Gotway, C. (1997). The effect of altering nitrogen and sulfur supply on the growth of cut chrysanthemums. *Journal of the American Society for Horticultural Science* **122**, 559–564.

Iida, A., Kazuoka, T., Torikai, S., Kikuchi, H. and Oeda, K. (2000). A zinc finger protein RHL41 mediates the light acclimatization response in *Arabidopsis*. *The Plant Journal* **24**, 191–203.

Ishikura, N. (1973). The changes in anthocyanin and chlorophyll content during the autumnal reddening of leaves. *Kumamoto Journal of Science, Biology* **11**, 43–50.

Jeannette, E., Reyss, A., Grégory, N., Gantet, P. and Prioul, J.-L. (2000). Carbohydrate metabolism in a heat-girdled maize source leaf. *Plant, Cell and Environment* **23**, 61–69.

Ji, S.-B., Yokoi, M., Saito, N. and Mao, L.-S. (1992). Distribution of anthocyanins in Aceraceae leaves. *Biochemical Systematics and Ecology* **20**, 771–781.

Johnson, J. D., Byres, D. P. and Dean, T. J. (1995). Diurnal water relations and gas exchange of two slash pine (*Pinus elliottii*) families exposed to chronic ozone levels and acidic rain. *New Phytologist* **131**, 381–392.

Kakegawa, K., Kaneko, Y., Hattori, E., Koike, K. and Takeda, K. (1987). Cell cultures of *Centaurea cyanus* produce malonated anthocyanin in UV light. *Phytochemistry* **26**, 2261–2263.

Kaku, S., Iwaya-Inoue, M. and Toki, K. (1992). Anthocyanin influence on water proton NMR relaxation times and water contents in leaves of evergreen woody plants during the winter. *Plant and Cell Physiology* **33**, 131–137.

Kaliamoorthy, S. and Rao, A. S. (1994). Effect of salinity on anthocyanin accumulation in the root of maize. *Indian Journal of Plant Physiology* **37**, 169–170.

Kao, C., Lai, H. H. and Chang, H. J. (1973). Effects of nitrogen, phosphorus and potassium deficiency in Chinese fir seedlings. *Memoirs of the College of Agriculture National Taiwan University* **14**, 86–92.

Kemmer, C. and Fritzsche, K. (1961). The effect of different nutrient levels and degrees of acidity on the variability of some morphological characters of one-year-old poplar plants. *Wissenschaftliche Abhandlungen/Deutsche Akademie der Landwirtschaftswissenschaften zu Berlin* **52**, 23–35.

Kennedy, B. F. and DeFilippis, L. F. (1999). Physiological and oxidative response to NaCl of the salt tolerant *Grevillea ilicifolia* and the salt sensitive *Grevillea arenaria*. *Journal of Plant Physiology* **155**, 746–754.

Knox, G. W. (1989). Water use and average growth index of five species of container grown woody landscape plants. *Journal of Environmental Horticulture* **7**, 136–139.

Krol, M., Gray, G. R., Hurry, V. M., Öquist, G., Malek, L. and Huner, N. P. A. (1995). Low-temperature stress and photoperiod affect an increased tolerance to photoinhibition in *Pinus banksiana* seedlings. *Canadian Journal of Botany* **73**, 1119–1127.

Lacey, C. J., Leaf, A. L. and Talli, A. R. (1966). Growth and nutrient uptake by flooded gum seedlings subjected to various phosphous supplies. *Australian Forestry* **30**, 212–222.

Landolt, W., Gunthardt-Goerg, M. S., Pfenninger, I and Scheidegger, C. (1994). Ozone-induced microscopical changes and quantitative carbohydrate contents of hybrid poplar (*Populus X euramericana*). *Trees: Structure and Function* **8**, 183–190.

Lee, D. W. and Collins, T. M. (2001). Phylogenetic and ontogenetic influences on the distribution of anthocyanins and betacyanins in leaves of tropical plants. *International Journal of Plant Science* **162**, 1141–1153.

Lee, D. W. and Lowry, J. B. (1980). Young-leaf anthocyanin and solar ultraviolet. *Biotropica* **12**, 75–76.

Leng, P., Itamura, H., Yamamura, H. and Deng, X. M. (2000). Anthocyanin accumulation in apple and peach shoots during cold acclimation. *Scientia Horticulturae* **83**, 43–50.

Li, W.-L., Berlyn, G. P. and Ashton, P. M. S. (1996). Polyploids and their structural and physiological characteristics relative to water deficit in *Betula papyrifera* (Betualceae). *American Journal of Botany* **83**, 15–20.

Maier-Maercker, U. (1997). Experiments on the water balance of individual attached twigs of *Picea abies* (L.) Karst. in pure and ozone-enriched air. *Trees: Structure and Function* **11**, 229–239.

Matile, P. (2000). Biochemistry of Indian summer: physiology of autumnal leaf coloration. *Experimental Gerontology* **35**, 145–158.

McKown, R., Kuroki, G. and Warren, G. (1996). Cold responses of *Arabidopsis* mutants impaired in freezing tolerance. *Journal of Experimental Botany* **47**, 1919–1925.

McMurtrey, J. E. Jr (1938). Distinctive plant symptoms caused by deficiency of any one of the chemical elements essential for normal development. *Botanical Review* **4**, 183–203.

Mehlenbacher, S. A. and Thompson, M. M. (1991). Inheritance of a chlorophyll deficiency in hazelnut. *HortScience* **26**, 1414–1416.

Michalk, D. L. and Huang, Z. K. (1992). Response of subterranean clover (*Trifolium subterraneum*) to lime, magnesium, and boron on acid infertile soil in subtropical China. *Fertility Research* **32**, 249–257.

Millar, A. A., Wrischer, M. and Kunst, L. (1998). Accumulation of very-long-chain fatty acids in membrane glycerolipids is associated with dramatic alterations in plant morphology. *The Plant Cell* **11**, 1889–1902.

Mills, T. M., Behboudian, M. H., Tan, P. Y. and Clothier, B. E. (1994). Plant water status and fruit quality in 'Braeburn' apples. *HortScience* **29**, 1274–1278.

Mita, S., Murano, N., Akaike, M., and Nakamura, K. (1997). Mutants of *Arabidopsis thaliana* with pleiotropic effects on the expression of the gene for beta-amylase and on the accumulation of anthocyanin that are inducible by sugars. *Plant Journal* **11**, 841–851.

Moran, J. A. and Moran, A. J. (1998). Foliar reflectance and vector analysis reveal nutrient stress in prey-deprived pitcher plants (*Nepenthes rafflesiana*). *International Journal of Plant Science* **159**, 996–1001.

Murray, J. R. and Hackett, W. P. (1991). Dihydroflavonol reductase activity in relation to differential anthocyanin accumulation in juvenile and mature phase *Hedera helix* L. *Plant Physiology* **97**, 343–351.

Murray, J. R., Smith, A. G. and Hackett, W. P. (1994). Differential dihydroflavonol reductase transcription and anthocyanin pigmentation in the juvenile and mature phases of ivy (*Hedera helix* L.). *Planta* **194**, 102–109.

Nii, N., Watanabe, T., Yamaguchi, K. and Nishimura, M. (1995). Changes of anatomical features, photosynthesis and ribulose bisphosphate carboxylase-oxygenase content of mango leaves. *Annals of Botany* **76**, 649–656.

Nouchi, I. and Odaira, T. (1973). Influence of ozone on plant pigments. *Taiki Osen Kenkyu* **8**, 120–125. (Translated by Translations, Environment Canada, 1981, No. 2024, 17 pp.)

Nozzolillo, C., Isabelle, P. and Das, G. (1990). Seasonal changes in the phenolic constituents of jack pine seedlings (*Pinus banksiana*) in relation to the purpling phenomenon. *Canadian Journal of Botany* **68**, 2010–2017.

Nygaard, P. H. (1994). Effects of ozone on *Vaccinium myrtillus*, *Hylocomium splendens*, *Pleurozium schreberi* and *Dicranum polysetum*. *Rapport fra Skogforskning* **9**, 1–17.

Opler, P. A., Frankie, G. W. and Baker, H. G. (1980). Comparative phonological studies of treelet and shrub species in tropical wet and dry forest in the lowlands of Costa Rica. *Journal of Ecology* **68**, 167–188.

Oren-Shamir, M. and Levi-Nissim, A. (1997a). Temperature effect on the leaf pigmentation of *Cotinus coggygria* 'Royal Purple'. *Journal of Horticultural Science* **72**, 425–432.

Oren-Shamir, M. and Levi-Nissim, A. (1997b). UV-light effect on the leaf pigmentation of *Cotinus coggygria* 'Royal Purple'. *Sciencia Horticulturae* **71**, 59–66.

Oswin, S. D., Kathieresan, K. and Deiva-Oswin, S. (1994). Pigments in mangrove species of Pichavaram. *Indian Journal of Marine Sciences* **23**, 64–66.

Paine, T. D., Hanlon, C. C., Pittenger, D. R. Ferrin, D. M. and Malinoski, M. K. (1992). Consequences of water and nitrogen management on growth and aesthetic quality of drought-tolerant woody landscape plants. *Journal of Environmental Horticulture* **10**, 94–99.

Pandolfini, T., Gabbrielli, R. and Ciscato, M. (1996). Nickel toxicity in two durum wheat cultivars differing in drought sensitivity. *Journal of Plant Nutrition* **19**, 1611–1627.

Parker, J. (1962). Relationships among cold hardiness, water-soluble protein, anthocyanins, and free sugars in *Hedera helix* L. *Plant Physiology* **37**, 809–813.

Rajendran, L., Ravishankar, G. A., Venkataraman, L. V. and Prathiba, K. R. (1992). Anthocyanin production in callus cultures of *Daucus carota* as influence by nutrient stress and osmoticum. *Biotechnology Letters* **14**, 707–712.

Ramanjulu, S., Veeranjaneyulu, K. and Sudhakar, C. (1993). Physiological changes induced by NaCl in mulberry var. Mysore local. *Indian Journal of Plant Physiology* **36**, 273–275.

Robinson, T. (1991). "The Organic Constituents of Higher Plants". Cordus Press, North Amherst, MA, USA.

Sakamoto, K., Kumiko, I., Sawamura, K., Kyoko, H., Yoshihisa, A., Takafumi, Y. and Tsutomu, F. (1994). Anthocyanin production in cultured cells of *Aralia cordata* Thumb. *Plant, Cell Tissue,and Organ Culture* **36**, 21–26.

Sakata, M. (1996). Evaluation of possible causes for the decline of Japanese cedar (*Cryptomeria japonica*) based on elemental composition and delta13C of needles. *Environmental Science and Technology* **30**, 2376–2381.

Schemske, D. W. and Bierzychudek, P. (2001). Evolution of flower color in the desert annual *Linanthus parryae*: Wright revisited. *Evolution* **55**, 1269–1282.

Schmucker, T. (1947). Anthocyanin im Holz der Rotbuche. *Naturwissenschaften* **34**, 91.

Sharma, P. N. (1995). Water relations and photosynthesis in phosphorus deficient mulberry plants. *Indian Journal of Plant Physiology* **38**, 298–300.

Sherwin, H. W. and Farrant, J. M. (1998). Protection mechanisms against excess light in the resurrection plants *Craterostigma wilmsii* and *Xerophyta viscosa*. *Plant Growth Regulation* **24**, 203–210.

Shirley, B. W., Kubasek, W. L., Storz, G., Bruggemann, E., Koornneef, M., Ausubel, F. M. and Goodman, H. M. (1995). Analysis of *Arabidopsis* mutants deficient in flavonoid biosynthesis. *Plant Journal* **8**, 659–671.

Sjulin, T. M., Mowrey, B. D., Amorao, A. Q., Coss, J. F., Nishimori, K. and Gilford, K. L. (2000) Strawberry plant named 'Alta Vista'. *US Patent Plant* **11,544**, 4 pp.

Spehar, C. R. and Galwey, N. W. (1996). Diallel analysis for aluminium tolerance in tropical soybeans [*Glycine max* (L.) Merrill]. *Theoretical and Applied Genetics* **92**, 267–272.

Spyropoulos, C. G. and Mavrommatis, M. (1978). Effect of water stress on pigment formation in *Quercus* species. *Journal of Experimental Botany* **29**, 473–477.

Suvarnalatha, G., Rajendran, L. and Ravishankar, G. A. (1994). Elicitation of anthocyanin production in cell cultures of carrot (*Daucus carota* L) by using elicitors and abiotic stress. *Biotechnology Letters* **16**, 1275–1280.

Suzuki, M. (1995). Enhancement of anthocyanin accumulation by high osmotic stress and low pH in grape cells (*Vitis* hybrids). *Journal of Plant Physiology* **147**, 152–155.

Tholakalabavi, A., Zwiabek, J. J. and Thorpe, R. A. (1994). Effect of mannitol and glucose-induced osmotic stress on growth, water relations, and solute composition of cell suspension cultures of poplar (*Populus deltoides* var *occidentalis*) in relation to anthocyanin accumulation. *In Vitro Cell Developmental Biology* **30P**, 164–170.

Tholakalabavi, A., Zwiazek, J. J. and Thorpe, T. A. (1997). Osmotically-stressed poplar cell cultures: anthocyanin accumulation, deaminase activity, and solute composition. *Journal of Plant Physiology* **151**, 489–496.

Thomas, D. A. and Barber, H. N. (1974). Studies on leaf characteristics of a cline of *Eucalyptus unigera* from Mount Wellington, Tasmania. I. Water repellency and the freezing of leaves. *Australian Journal of Botany* **22**, 501–512.

Tignor, M. E., Davies, F. S., Sherman, W. B. and Davis, J. M. (1997). Rapid freezing acclimation of *Poncirus trifoliata* seedlings exposed to 10°C and long days. *HortScience* **32**, 854–857.

Tuohy, J. M. and Choinski, J. S. Jr (1990). Comparative photosynthesis in developing leaves of *Brachystegia spiciformis* Benth. *Journal of Experimental Botany* **41**, 919–923.

Van Huystee, R. B., Weiser, C. J. and Li, P. H. (1967). Cold acclimation in *Cornus stolonifera* under natural and controlled photoperiod and temperature. *Botanical Gazette* **128**, 200–205.

Vogt, T., Ibdah, M., Schmidt, J., Wray, V., Nimtz, M. and Strack, D. (1999). Light-induced betacyanin and flavonol accumulation in bladder cells of *Mesembryanthemum crystallinum*. *Phytochemistry* **52**, 583–592.

Wallace, I. M., Dell, B. and Loneragan, J. F. (1986). Zinc nutrition of jarrah (*Eucalyptus marginata* Donn ex Smith) seedlings. *Australian Journal of Botany* **34**, 41–51.

Weiss, D. (2000). Regulation of flower pigmentation and growth: multiple signaling pathways control anthocyanin synthesis in expanding petals. *Physiologia Plantarum* **110**, 152–157.

Wettstein-Westersheim, W. and Minelli, H. (1962). Breeding inter-sectional Poplar hybrids. *Allgemeine Forstzeitung* **73S**, 2.

Whatley, J. M. (1992). Plastid development in distinctively coloured juvenile leaves. *New Phytologist* **120**, 417–426.

Wood, B. W., Grauke, L. J. and Payne J. A. (1998). Provenance variation in pecan. *Journal of the American Society for Horticultural Science* **123**, 1023–1028.

Woodall, G. S., Dodd, I. C. and Stewart, G. R. (1998). Contrasting leaf development within the genus *Syzygium*. *Journal of Experimental Botany* **49**, 79–87.

Wright, J. W. (1944). Genotypic variation in white ash. *Journal of Forestry* **42**, 489–495.

Xiang, C., Werner, B. L., Christensen, E. M. and Oliver, D. J. (2001). The biological functions of glutathione revisited in *Arabidopsis* transgenic plants with altered glutathione levels. *Plant Physiology* **126**, 564–574.

Zakhleniuk, O. V., Raines, C. A. and Lloyd, J. C. (2001). pho3: a phosphorus-deficient mutant of *Arabidopsis thaliana* (L.) Heynh. *Planta* **212**, 529–534.

Zhi-min, Y., Shao-jian, Z., Ai-tong, H., You-fei, Z. and Jing-yi, Y. (2000). Response of cucumber plants to increased UV-B radiation under water stress. *Journal of Environmental Sciences* **12**, 236–240.

The Role of Anthocyanins for Photosynthesis of Alaskan Arctic Evergreens During Snowmelt

STEVEN F. OBERBAUER[1] and GREGORY STARR[1,2]

[1]*Department of Biological Sciences, Florida International University, Miami, FL 33199, USA*
[2]*School of Forestry Resources and Conservation, University of Florida, Gainesville, FL 32611, USA*

ABSTRACT

The period during spring snowmelt represents a challenging time for arctic tundra evergreens. As a result of relatively clear skies and increasing solar zeniths, radiation loads can be high, even under snow cover, while air and soil temperatures may be very low. Anthocyanin pigmentation is a prominent feature of most tundra evergreens during the spring melt. We suggest that these anthocyanins protect plants from photodamage at this critical period. The patterns of anthocyanin production and occurrence in tundra evergreen plants are compatible with light screening by these pigments. Anthocyanins are located in the outer palisade cells, consistent with protection of the photosynthetic tissues from radiation. Anthocyanin concentrations increase while plants remain under snow cover and attain their maximum shortly after snowmelt in the spring, when radiation levels are highest but temperatures remain low. Photosynthetic activity is present under the snow in tundra evergreen plants, and is similar in pattern to anthocyanin concentrations, increases as depth of snow cover declines. Anthocyanin content declines once temperatures increase during the growing season, but increase again with the occurrence of fall frost at the end of the growing season. Variation in anthocynanin pigmentation is related to plant microsite (exposure to light) and leaf cohort; leaves with high nitrogen (exposed or young leaves) have higher anthocyanin concentrations. We believe this light screening strategy is an important aspect of the acquisition of carbon reserves for many tundra evergreen plants during the short growing season.

Advances in Botanical Research Vol. 37
incorporating Advances in Plant Pathology
ISBN 0-12-005937-1

I. INTRODUCTION

A. CHALLENGES TO PLANTS DURING SNOWMELT

Until recently, research on the physiology of vascular plants in arctic tundra has focused on the short summer growing season on the assumption that conditions during the remainder of the year were too harsh to allow for any significant leaf physiological activity (Tieszen, 1981). However, striking changes in anthocyanin pigmentation in many arctic evergreens observed over winter and spring and evidence for a photoprotective function of these pigments (Gould *et al.*, 1995, 2000; Feild *et al.*, 2001; Hoch *et al.*, 2001) suggest that important physiological processes are taking place at these times (Grogan *et al.*, 2001; Starr and Oberbauer, unpublished data). Several recent studies have provided strong evidence that ecosystem level processes detected above snow cover during the winter and spring in the Arctic are a result of the active physiology of vascular plants under snow cover (Oechel *et al.*, 2000; Grogan *et al.*, 2001; Starr and Oberbauer, unpublished data). Anthocyanins may play an important role in these processes during this critical period of the year.

The growing season in the Alaskan Arctic is typically short, lasting only 8–12 weeks during which plants must accumulate sufficient carbon to provide for growth, reproduction, and maintenance. Any mechanisms that maximize carbon gain during this period are likely to be advantageous. Much of the year outside the growing season is characterized by little or no sunlight and temperatures substantially below freezing (Chapin and Shaver, 1985). However, for several weeks in the spring and fall, irradiance levels may be quite high, while temperatures are highly variable; moderate temperatures above freezing may be interspersed with hard freezes. In May and early June, snow cover often persists long after air temperatures and light levels would be suitable for photosynthesis of vascular plants without snow cover. Timing of snowmelt, however, is

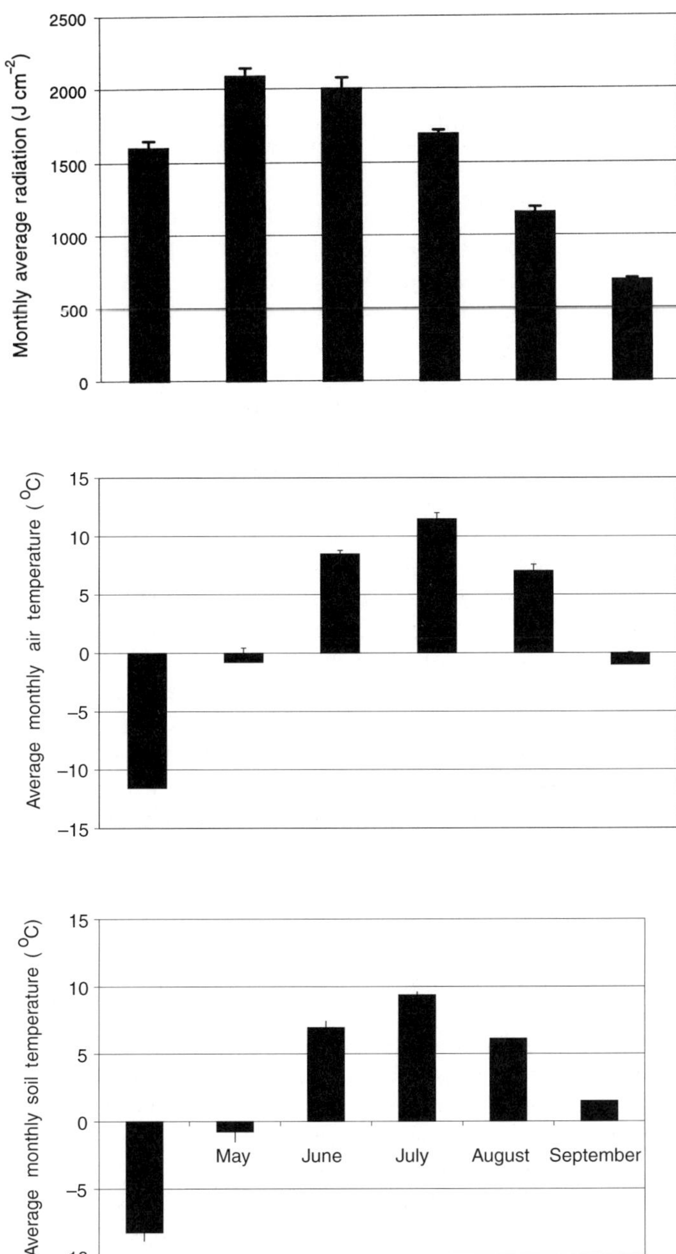

Fig. 1. Monthly mean global radiation (shortwave, J cm^{-2}) for 1990–2000 (upper) and mean monthly air (middle) and soil temperature at 5 cm depth (lower) at Toolik Field Station Long Term Ecological Research weather station.

highly variable and depends on both snow depth and density, arrival of warm air masses, and prevalence of clear skies. Snow 20 cm deep can melt in a few days or persist for weeks. Date of snowmelt at Toolik Lake, Alaska over the last 7 years has varied by 26 d with the earliest date on 13 May and the latest 9 June. At Imnavait Creek, a site close to Toolik, full snow melt ranged from 12 May to 11 June over a 17 year record (Kane *et al.*, 2000).

The period during snowmelt represents a challenging time for tundra evergreens. As a result of relative clear skies and increasing solar zeniths, radiation loads can be very high. At Toolik, the highest radiation loads of the year occur in May (Fig. 1). Because significant radiation penetrates the snow (Hamerlynck and Smith, 1994; Liston *et al.*, 1999; Starr and Oberbauer, unpublished data), evergreen plants are exposed to light levels that can be either potentially damaging or useful. As snow melts and plants project above the snow, radiation levels on leaves can be extreme due to reflectance off the surface of the snow (DeLucia *et al.*, 1991). At the same time, plants may experience relatively low air temperatures, and equally important, soil temperatures below freezing (Fig. 1). Low temperatures inhibit photosynthesis both by increasing the CO_2 diffusion resistances within leaves when solutes freeze (~ –5°C) and by directly affecting photosynthetic membrane and enzyme function (Kappen, 1993; Larcher, 1995). Low soil temperatures increase root resistance (Kramer, 1940) and water viscosity (Nobel, 1999). The combination of high irradiances and low temperatures are known to cause significant photoinhibition and photodamage to leaves exposed to the combination of these conditions (Hamerlynck and Smith, 1994; Manuel *et al.*, 1999; Germino and Smith, 2000; Gould *et al.*, 2000).

Evergreen plants under the snow and during snow melt have several alternatives for coping with the high radiation loads (Streb *et al.*, 1998; Manuel *et al.*, 1999; Germino and Smith, 2000): use the light for carbon fixation; minimize absorption by increased reflectance or minimizing cross-sectional area exposed; screen the high irradiance with photoprotective pigments such as anthocyanins; and dissipate the energy via photochemical and non-photochemical means. A combination of these processes is likely occurring in tundra evergreens during snowmelt.

B. ANTHOCYANINS IN TUNDRA EVERGREEN PLANTS

The presence of leaf anthocyanins is extremely widespread among tundra plants, with spectacular fall coloration in deciduous species the norm. However, anthocyanins are also widespread in evergreen tundra species in the Ericaceae or closely related families (Plate 5, Table I). Notable

TABLE I
Common tundra evergreens at Toolik Lake, Alaska showing strong seasonal variation in anthocynanin pigmentation. Nomenclature follows that of Hultén (1968)

Species	Family	Growth form
Andromeda polifolia L.	Ericaceae	Dwarf erect shrub
Cassiope tetragona (L). D. Don	Ericaceae	Dwarf erect shrub
Diapensia lapponicum L.	Ericaceae	Cushion plant
Empetrum nigrum L.	Ericaceae	Dwarf creeping shrub
Ledum palustre (L.)	Ericaceae	Dwarf erect shrub
Oxycoccus microcarpus Turcz.	Ericaceae	Filiform prostrate
Rhododendron lapponicum (L.) Wahlenb.	Ericaceae	Dwarf erect shrub
Vaccinium vitis-idaea L.	Ericaceae	Dwarf creeping shrub
Pyrola grandiflora Radius	Pyrolaceae	Rosette
Dryas integrifolia M. Vahl.	Rosaceae	Dwarf creeping shrub

exceptions are the wintergreen graminoids, including the dominant species in tussock tundra, *Eriophorum vaginatum*. In four species examined (*Cassiope tetragona*, *Empetrum nigrum*, *Ledum palustre*, and *Vaccinium vitis-idaea*), anthocyanins were found in palisade cells just beneath the outer epidermis (Plate 5).

The hypothesis underlying this study is that anthocyanins play an important role in photoprotection of photosynthetic apparatus in tundra evergreen species under the high irradiances and low temperatures during snow melt. This hypothesis, 'the light screen hypothesis', suggests that anthocyanins protect the photosynthetic apparatus from photoinhibition when high light is coupled with low air temperatures (Feild *et al.*, 2001; Hoch *et al.*, 2001). The light screen hypothesis engenders a number of testable predictions. These are that anthocyanins should be:

1. located in leaf tissues so as to encounter incoming radiation before it is intercepted by the bulk of the photosynthetic cells;
2. at their highest concentrations when radiation levels are highest and temperatures are low;
3. highest in tissues with the greatest investment in the photosynthetic apparatus.

Although this hypothesis has been evaluated for leaves of temperate species during autumn senescence, this is the first evaluation of the hypothesis for arctic evergreen species during the spring snowmelt period. The intent of this paper is to review recent findings on the role of anthocyanins for photosynthesis of arctic evergreens during the snow melt season. The majority of the findings come from studies conducted on evergreen plants near Toolik, Alaska (68° 38′N, 149° 34′W, elevation 730 m; Walker *et al.*, 1994) during May of 1997, 1998, 1999, 2000 and 2001.

II. EVIDENCE FOR PHOTOPROTECTION BY ANTHOCYANINS

A. SUBNIVIAN MICROCLIMATE

As melt-out approaches, springtime snow cover creates an environment that is favorable for photosynthesis of plants under the snow. The insulative properties of snow buffer plant temperatures so they usually remain above –5.0°C at this time of year despite air temperatures that may be substantially lower. Midday temperatures of monitored leaves approached or exceeded 0°C. Spot measurements of temperature in the air spaces and at the moss/soil surface within the depth hoar can be 5.0°C warmer than air temperatures above the snow (Starr and Oberbauer, unpublished data). Sufficient light penetrates beneath the snow to drive photosynthesis at depths less than 30 cm (Fig. 2). At snow depths of 28 cm, photosynthetic photon flux density (PPFD) can reach 300 μmol m^2 s^{-1} and increase to 800 μmol m^2 s^{-1} at solar noon when snow depths are below 20 cm. At the same time, build up of respiratory CO_2 under the snow may enhance CO_2 uptake (Starr and Oberbauer, unpublished data). CO_2 concentrations within the depth hoar layer are substantially elevated over ambient atmospheric levels with average values ranging from 405 to 675 and individual readings exceeding 1000 μmol mol^{-1} (Fahnestock *et al.*, 1998; Starr and Oberbauer, unpublished data).

B. PHOTOSYNTHESIS DURING SNOWMELT

That tundra evergreen plants are photosynthetically competent upon emergence from the snow has been known for some time (Oberbauer *et al.*, 1996), but clear evidence that photosynthetic activity under the snow occurs in tundra evergreens has only recently been reported (Starr, 2000; Starr and Oberbauer, unpublished data). Significant photosystem II (PSII) activity and carbon uptake capacity has been found in leaves of four tundra species, *Eriophorum vaginatum*, *Ledum palustre*, *Vaccinium vitis-idaea* and *Cassiope tetragona*. Although the averages across all plants for both parameters were low, some individuals of all species showed high photosynthetic capacity. Rates as high as 35% of the species physiological maximum were recorded: 3.98, 1.42, 1.60, and 2.51 μmol CO_2 m^{-2} s^{-1} for *E. vaginatum*, *L. palustre, V. vitis-idaea*, and *C. tetragona* respectively. Rates under snow measured at saturating light (A_{max}) in spring 1999 were elevated relative to A_{amb} rates in 1998 (Table II). The mean A_{max} of three of the species was well above the physiological compensation point (Table II) and for some plants exceeded 60% of A_{max} at mid-season (Oberbauer and Oechel, 1989). *Ledum palustre* showed less photosynthetic activity, averaging near the compensation point (Table II). Snow cover depth and A_{max} were not significantly correlated for the four

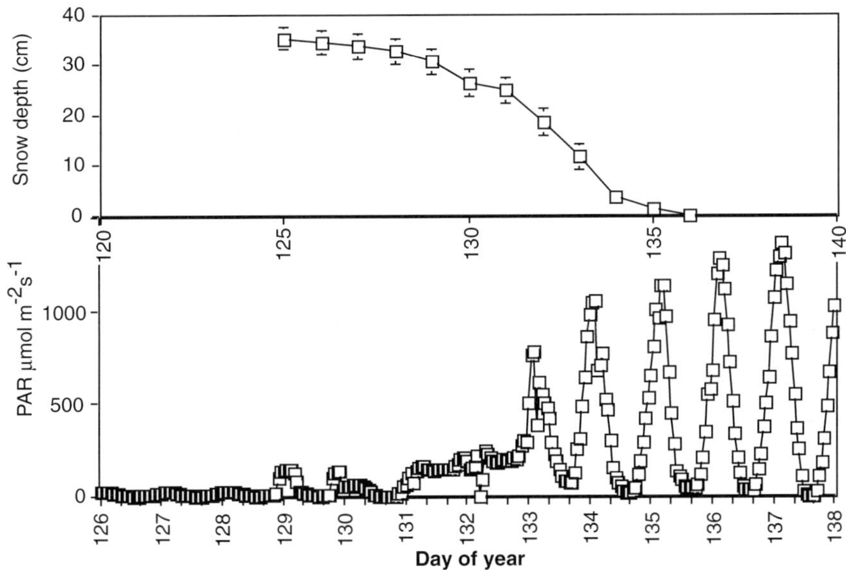

Fig. 2. Light penetration to the vegetation under the snow and snow depth during the spring of 1999, a year with very rapid snowmelt. Values are means of 19 sensors.

species (Table III), probably as a result of variation in temperature and CO_2 concentrations.

The average F_v/F_m of these four species was approximately half of their summer maximum (Table II, Figure 3). Similar to A_{max}, the variance in F_v/F_m was high: some individuals exceeded 90% of summer maximum. Unlike A_{max}, F_v/F_m increased as depth of snow cover decreased. This relationship was statistically significant for *E. vaginatum*, *L. palustre*, *V. vitis-idaea*, but not for *C. tetragona* (Table III).

All four species do not show equivalent photosynthetic activity under the snow. The lower photosynthetic capacity of *L. palustre* may be attributed in part to their leaf orientation. During winter the leaves of this species are strongly deflexed against the stem, reducing light absorption and possibly hindering the diffusion of CO_2 to the stomata (Smith *et al.*, 1997). One would also expect the CO_2 exchange of the other three species to be higher than measured because they maintained a relatively high level of PSII activity. Low soil temperatures under these conditions likely limit water uptake and induce stomatal closure (Delucia, 1986; Day *et al.*, 1989, 1990). Laboratory studies of *E. vaginatum* report stomatal closure in response to low soil temperatures independent of air temperature (Starr, Neuman and Oberbauer, unpublished data). Manuel *et al.* (1999) reported that the wintergreen alpine plant, *Geum montanum*, is not subject to photodamage at high light, low temperature conditions,

TABLE II

Mean photosynthesis (1 ± SE) under ambient light (A_{amb}, 1998) and saturating light (A_{max}, 1999) and F_v/F_m for evergreens under the snow. n = 50 measured in May of 1998 or 1999

	Ledum palustre	*Vaccinium vitis-idaea*	*Eriophorum vaginatum*	*Cassiope tetragona*
1998				
A_{amb} (μmol m^{-2} s^{-1})	–0.364 ± 0.157	0.032 ± 0.119	–0.066 ± 0.500	0.089 ± 0.411
F_v/F_m	0.294 ± 0.009	0.332 ± 0.007	0.448 ± 0.011	0.339 ± 0.012
1999				
A_{max} (μmol m^{-2} s^{-1})	–0.06 ± 0.16	1.05 ± 0.16	1.31 ± 0.65	1.11 ± 0.31
F_v/F_m	0.33 ± 0.02	0.39 ± 0.01	0.46 ± 0.02	0.40 ± 0.04

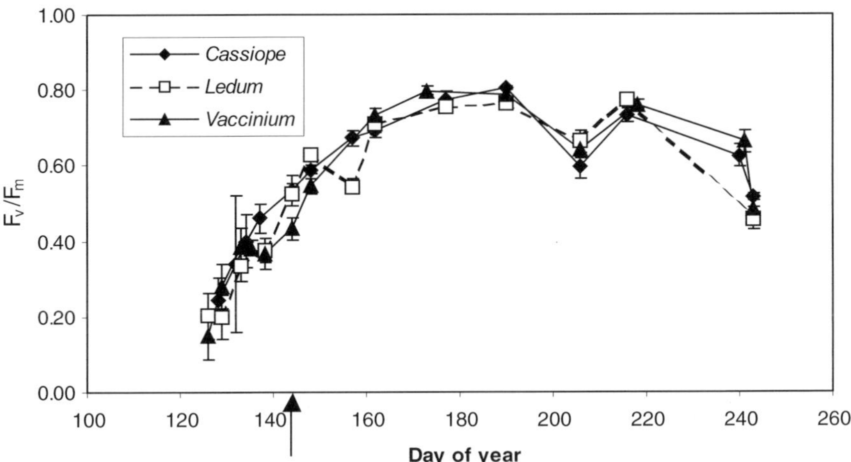

Fig. 3. Seasonal patterns of chlorophyll fluorescence (F_v/F_m, mean ± 1 SE) of three arctic evergreens in Alaskan tundra during the 1999 growing season at Toolik Lake, Alaska, $n = 10$. Arrow at day of year 142 indicates completion of snowmelt.

even under desiccation stress. This tolerance apparently results from an interplay between photorespiration and photoassimilation by a combination of mechanisms including shifting excitation energy from PSII to photosystem I (PSI), cyclic electron transport, and the xanthophyll cycle. Similar mechanisms are likely interacting in the arctic evergreens along with light screening by anthocyanins.

For tundra plants, the requirement for carbon acquisition is great after a long winter during which light input is effectively zero and foliar respiration accounts for large carbon losses. With these four species able to activate their photosynthesis under the snow during the spring, they partially compensate for their winter losses. This activity also enables the plants to reach their maximum photosynthetic capacity quickly once melt has occurred (Defoliart *et al.*, 1988; Kudo, 1999). Consequently subnivean photosynthesis appears to be an adaptive trait for survival of the species at high latitudes, where the growing season is limited (Huner *et al.*, 1998).

C. SEASONAL CHANGES IN PIGMENT CONCENTRATIONS

Of the four species examined, the sedge, *E. vaginatum*, did not contain any anthocyanins. The three species of ericaceae (*Cassiope*, *Vaccinium*, and *Ledum*) emerge from the snow in the spring with intense anthocyanin pigmentation (Starr and Oberbauer, unpublished data). Excavation of plants under the snow revealed that anthocynanin concentrations, while already elevated under deep snow, significantly increased as snow depth decreased in *L. palustre* and *V. vitis-idaea* (Table III). The same tendency was found for *C. tetragona*. In all three species, anthocyanin concentrations reach their maximum immediately following snowmelt, then rapidly decline during the short growing season (Fig. 4), a finding consistent with the predictions of the light screen hypothesis. In the early fall anthocyanin production increases again following the onset of fall frosts. Fall concentrations were much lower than those encountered during the spring season as snow melts. The limited production of pigments during the fall can be attributed to the relatively lower light levels at that time of the year: at Toolik, day length in August decreases approximately 18 min/d and cloud cover is typically high (Fig. 1).

Total carotenoid concentrations also increased with decreasing snow depth in three of the four species, but the increase was significant only for *C. tetragona* (Table III). In the exception, *V. vitis-idaea*, carotenoids significantly decreased as snowmelt proceeded in the spring. Most study species tended to increase both their chlorophyll a and chlorophyll b concentrations as snow depth declined, however *C. tetragona* was the only one to show significant changes during this period (Table III).

D. PHOTOSYNTHESIS AND PIGMENTATION

Based on the light screening hypothesis, a relationship between pigment concentrations and CO_2 exchange might be expected. Correlations between pigments concentrations and A_{max} were not significant, again probably because of variation in subnivean CO_2 concentrations and temperatures. However, significant correlations were found between potential quantum yield, F_v/F_m, and anthocyanin concentrations for *L. palustre* and *V. vitis-idaea* ($P<0.001$ for all cases), but not for *C. tetragona* (Figs 3 and 4). The same trend was also seen between total carotenoids

TABLE III
*Pearson correlation analyses of photosynthetic characteristics and pigment concentrations with snow depth. r-values are presented. * indicates significance at P<0.05 and ** at P<0.001*

	Ledum palustre	*Vaccinium vitis-idaea*	*Eriophorum vaginatum*	*Cassiope tetragona*
A_{max}	0.1726	0.0418	0.0327	0.2557
F_v/F_m	–0.4255**	–0.3937**	–0.4255**	–0.2504
Chlorophyll a	0.0730	–0.2161	–0.2013	–0.2162
Chlorophyll b	–0.0132	0.1679	–0.1286	–0.2868*
Total carotenoids	–0.0462	0.3110*	–0.2527	–0.3000*
Anthocyanins	–0.3123*	–0.4620**		–0.3000

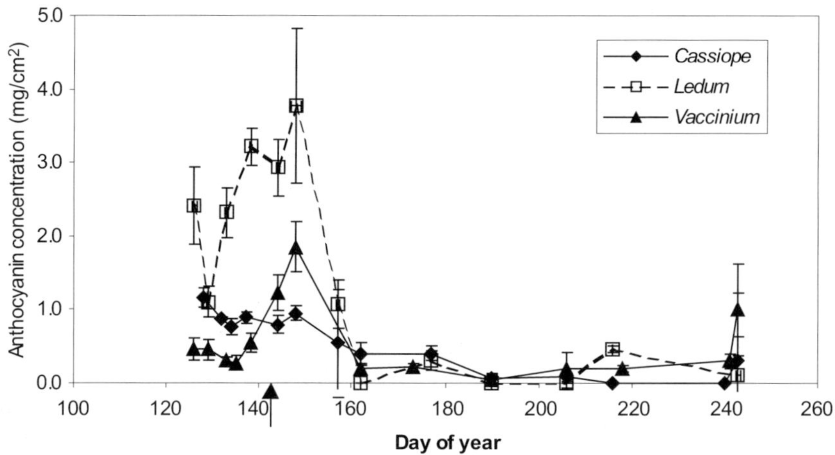

Fig. 4. Seasonal patterns of anthocyanin content (A/cm^2, mean $\pm$ 1 SE) of three arctic evergreens in Alaskan tundra during the 1999 growing season at Toolik Lake, Alaska, $n = 10$. Arrow at day of year 142 indicates completion of snowmelt.

and potential quantum yield in *L. palustre*, and *V. vitis-idaea* (P<0.001). These correlations are related to changes in snow depth. Deeper snow reduces the quantity of light reaching the plants while temperatures are still cold, therefore reducing the need for the protective pigments to be produced and allowing for PSII to be maintained (Streb *et al.*, 1998; Neuner *et al.*, 1999).

E. IMPORTANCE OF MICROSITE

In tussock tundra, the dominant species, *E. vaginatum*, produces raised tussocks of variable heights from 10 to 30 cm above the surrounding vegetation surface. Tundra dwarf shrubs measured in this study are most commonly found in the inter tussock spaces and on the sides of tussocks. As of result of this microtopographic variation, individual plants of a given species may experience very different micro-environmental conditions. These differences include light environment, as well as air temperatures and depth of snow cover. Dwarf evergreen shrubs, *Ledum*, *Cassiope*, and *Vaccinium* exposed to high light environments with little snow cover have been observed to have heavy anthocynanin pigmentation and lower photosynthetic capacity than bright green plants in depressions or on north sides of tussocks (Starr and Oberbauer, unpublished data). The light screen hypothesis predicts that plants in high light microsites should have the highest concentrations of anthocyanins during spring melt. Such circumstances would include plants located on the southern sides of tussocks, leaning toward south, or with a portion of the plant oriented into the direct light. In the case of *C. tetragona*, which typically has vertical stems with closely appressed leaves in four vertical rows, the leaves on the south side receive more light than those on the north side.

To evaluate these position effects, Starr and Oberbauer (unpublished data) evaluated plant orientation under the snow and as they emerged. Azimuth of the stem and stem angle had a significant effect on *C. tetragona*; plants; with leaves facing a southern direction had the highest concentrations of anthocyanins (Table IV). In addition leaves of this species which were orientated towards the sky were excavated from under the snow load and compared with those compressed to the tundra floor by the snow. In all cases, upwards facing leaves with potentially high irradiance exposures had significantly higher concentrations of anthocyanins (Table V).

Given that many of these species ranges span large distances, Starr and Oberbauer (unpublished data) sampled two species of evergreens for anthocyanin concentrations while under snow along a 320 km transect on the Dalton Highway in Alaska. Results were equivocal, with a nearly significant increase in anthocyanin for one of the two species at the more northerly sites (Table VI).

F. EFFECTS OF LEAF AGE

The three study species containing anthocyanin all have leaf life spans greater than one year (Reich *et al.*, 1997). *Ledum* leaves produced mid-June typically last for 1 to 2 years at the study site. Leaves of the other two species, *Cassiope* and *Vaccinium* typically persist for 4 or more years. In all these species, differences in anthocyanin content of the youngest versus older leaves are visually apparent, particularly during snowmelt. Pigment extraction confirmed that the youngest leaves of *Ledum* and *Vaccinium* have significantly greater anthocyanin concentrations than older leaf cohorts, a finding consistent with the light screen hypothesis (Table V).

Nitrogen is the primary limiting nutrient in tussock tundra, which is characterized by low nutrient availability resulting from low decomposi-

TABLE IV

*Multiple regression analyses to determine the role that position, angle of leaves and snow depth had on anthocyanin concentrations. n = 30. * indicates significance of P<0.05 and ** P<0.005*

	Vaccinium vitis-idaea	*Ledum palustre*	*Cassiope tetragona*
R^2	0.075	0.083	0.645
P-values	0.898	0.423	0.004**

TABLE V

*Comparison of anthocyanin concentration (A_{525}/cm^2) and nitrogen concentration (gN/g dry wt *100) in 1999 and 2000 leaf cohorts and sun vs. shade leaves using Kruskal-Wallis ANOVAS. n = 30. Values represent means ± SE*

	Vaccinium vitis-idaea		*Ledum palustre*		*Cassiope tetragona*	
Leaf cohort	1999	2000	1999	2000	Shade	Sun
Anthocyanin (A_{525}/cm^2)	0.048 ± 0.099	0.979 ± 0.163	0.594 ± 0.316	1.660 ± 0.285	0.085 ± 0.024	0.244 ± 0.085
P-value	0.0001**		0.0001**		0.0500*	
% Nitrogen (g/g*100)	0.8 ± 0.03	0.93 ± 0.03	1.4 ± 0.03	1.6 ± 0.05	0.95 ± 0.03	1.05 ± 0.02
P-value	0.01**		0.01**		0.10	

TABLE VI

Comparisons of anthocyanin concentrations for evergreens under the snow along a south–north Alaskan tundra transect. P-values indicate significance determined by Kruskal-Wallis ANOVAS. n = 8. Absorbance (A/cm²) ± 1 SE

	Ledum palustre	*Vaccinium vitis-idaea*
South		
Gobler's Nob	0.6264 ± 0.129	0.6964 ± 0.198
Finger Mountain	1.5800 ± 0.237	0.8968 ± 0.171
↓ Toolik	1.4480 ± 0.226	0.5798 ± 0.208
North		
P-values	0.0907	0.8529

tion and nitrogen mineralization rates (Nadelhoffer *et al.*, 1992). Arctic species efficiently translocate limited nutrients into new leaves, and young leaves typically have higher nitrogen contents than older leaf cohorts (Shaver *et al.*, 1992; Starr *et al.*, 2000). Leaf cohorts produced later in the growing season will have a high requirement for carbon acquisition to balance the plant carbon budget. The high cost of producing an evergreen leaf in this inherently nutrient-poor environment means that protection of the leaves with the highest payback time and nitrogen concentration would be the most cost effective (Wookey *et al.*, 1994; Kikuzawa, 1995; Kudo, 1999). Interestingly, during mid and late season when older leaf cohorts are senescing and nutrients are being retranslocated, they typically have stronger anthocyanin pigmentation than younger leaves. Consistent with the light screen hypothesis, development of anthocyanin in the fall for all leaf cohorts can be inhibited by artificial shading (Oberbauer, unpublished data).

III. CONCLUSIONS AND PERSPECTIVES

Anthocyanin pigmentation is a prominent feature in most tundra evergreens during the fall and spring. We believe that these anthocyanins protect plants from photodamage when radiation levels are high and temperatures can be very low. The patterns of anthocyanin production and occurrence in tundra evergreen plants are consistent with this light screening hypothesis. The cellular location of anthocyanins is consistent with protection of the photosynthetic tissues from radiation. Anthocyanin concentrations increase while plants remain under snow cover and reach their maximum shortly after snowmelt in the spring, when radiation levels are high but temperatures are still low. Photosynthetic activity is present under the snow in these plants, and similar to anthocyanin concentrations, increases as depth of snow cover declines. Anthocyanin content declines once temperatures increase during the growing season, but increases again with the occurrence of fall frost at the end of the growing season. Variation in anthocyanin pigmentation is related to plant microsite (exposure to light) and leaf cohort; leaves with high nitrogen (exposed leaves or younger) have higher anthocyanin concentrations.

Much remains to be learned about the function of anthocyanins in tundra evergreens. An obvious need is more detailed study of the xanthophyll cycles and anthocyanin light screening. The photosynthetic reflectivity index (PRI) would be an appropriate first approach (Gamon *et al.*, 1992; Pinuelas *et al.*, 1995; Gamon and Surfus, 1999). In particular the role of the xanthophyll cycle in non-anthocyanin containing evergreens such as *Eriophorum vaginatum* would be particularly interesting. Another promising direction is the interaction between light, temperature and daylength for the production of anthocyanins. Anthocyanin production is induced with declining photoperiod in some of these species (Oberbauer, unpublished data).

Understanding the role of anthocyanins in arctic evergreens is becoming increasingly important as the evidence is now overwhelming that the climate is warming in the Arctic (IPPC, 2001). Warming is expected to alter the amount and duration of snow cover (Maxwell, 1992). How tundra evergreens respond to changes in snow cover and snowmelt is likely going to be related to the functioning of anthocyanin in these plants.

ACKNOWLEDGEMENTS

This study was supported by the National Science Foundation Office of Polar Programs grants OPP-9615845 and OPP-9907185 as part of the International Tundra Experiment Program (ITEX).

REFERENCES

Chapin, F. S. III and Shaver, G. R. (1985). Arctic. *In* "Physiological Ecology of North American Plant Communities" (B. F. Chabot and H. A. Mooney, eds), pp. 16–40, Chapman and Hall, New York.

Day, T. A., DeLucia, E. H. and Smith, W. K. (1989). Influence of cold soils and snowcover on photosynthesis and leaf conductance in two Rocky Mountain conifers. *Oecologia* **80**, 546–552.

Day, T. A., DeLucia, E. H. and Smith, W. K. (1990). Effect of soil temperature on stem sap flow, shoot gas exchange and water potential of *Picea engelmannii* Parry during snowmelt. *Oecologia* **84**, 474–481.

Defoliart, L. S., Griffith, M., Chapin, F. S. III and Jonasson, S. (1998). Seasonal patterns of photosynthesis and nutrient storage in *Eriophorum vaginatum* L., and arctic sedge. *Functional Ecology* **2**, 185–194.

DeLucia, E. H. (1986). Effect of low root temperature on net photosynthesis, stomatal conductance and carbohydrate concentration in Engelmann spruce (*Picea engelmannii* Parry ex Engelm.) seedlings. *Tree Physiology* **2**, 143–154.

DeLucia, E. H., Day, T. A. and Öquist G. (1991). The potential for photoinhibition of *Pinus silvestris* L. seedlings exposed to high light and low soil temperatures. *Journal of Experimental Botany* **42**, 611–617.

Fahnestock, J. T., Jones, M. H., Brooks, P. D., Walker, D. A. and Welker, J. M. (1998) Winter and early spring CO_2 efflux from tundra communities of northern Alaska. *Journal of Geophysical Research* **103**, 29 023–29 027.

Feild, T. S., Lee, D. W. and Holbrook, N. M. (2001). Why leaves turn red in autumn. The role of anthocyanins in senescing leaves of red-osier dogwood. *Plant Physiology* **127**, 566–574.

Gamon, J. A. and Surfus, J. S. (1999). Assessing leaf pigment content and activity with a reflectometer. *New Phytologist* **143**, 105–117.

Gamon, J. A., Penuelas, J. and Field, C. B. (1992). A narrow-waveband spectral index that tracks diurnal changes in photosynthetic efficiency. *Remote Sensing of the Environment* **41**, 35–44.

Germino, M. J. and Smith, W. K. (2000). Differences in microsite, plant form, and low-temperature photoinhibition in alpine plants. *Arctic, Antarctic, and Alpine Research* **32**, 388–396.

Gould, K. S., Kuhn, D. N., Lee, D. W. and Oberbauer, S. F. (1995). Why leaves are sometimes red. *Nature* **378**, 241–242.

Gould, K. S., Markham, K. R., Smith, R. H. and Goris, J. (2000). Functional role of anthocyanins in the leaves of *Quintinia serrata* A. Cunn. *Journal of Experimental Botany* **51**, 1107–1115.

Grogan, P. , Illeris, L. , Michelsen, A. and Jonasson, S. (2001). Respiration of recently-fixed plant carbon dominates mid-winter ecosystem CO_2 production in sub-arctic heath tundra. *Climatic Change* **50**, 129–142.

Hamerlynck, E. P. and Smith, W. K. (1994) Subnivean and emergent microclimate, photosynthesis, and growth in *Erythronium grandiflorum* Pursh, a snowbank geophyte. *Arctic and Alpine Research* **26**, 21–28.

Hoch, W. A., Zeldin, E. L. and McCown, B. H. (2001). Physiological significance of anthocyanins during autumnal leaf senescence. *Tree Physiology* **21**, 1–8.

Hultén, E. (1968). "Flora of Alaska and Neighboring Territories". Stanford University Press, Stanford.

Huner, N. P. A., Öquist, G. and Sarhan, F. (1998). Energy balance and acclimation to light and cold. *Trends in Plant Sciences* **3**, 224–230.

Intergovernmental Panal on Climate Change (IPCC) (2001). "Summary for Policy Makers: The Third Assessment Report of Working Group I of the Intergovernmental Panel on Climate Change". Cambridge University Press, Cambridge.

Kane, D. L., Hinzman, L. D., McNamara, J. P., Zhang, Z. and Benson, C. S. (2000). An overview of a nested watershed study in Arctic Alaska. *Nordic Hydrology* **31**, 245–266.

Kappen, L. (1993). Plant activity under snow and ice, with particular reference to lichens. *Arctic* **46**, 297–302.

Kikuzawa, K. (1995). Leaf phenology as an optimal strategy for carbon gain in plants. *Canadian Journal of Botany* **73**, 158–163.

Kramer, P. J. (1940). Root resistance as a cause of decreased water absorption at low soil temperatures. *Plant Physiology* **16**, 63–79.

Kudo, G. (1999). A review of ecological studies on leaf-trait variations along environmental gradients – in the case of tundra plants. *Japanese Journal of Ecology* **49**, 21–35.

Larcher, W. (1995). "Physiological Plant Ecology". Springer Verlag, Berlin.

Liston, G. E., Wintherm, J. G., Bruland, O., Elvehoy, H. and Sand, K. (1999). Below-surface ice melt on the coastal Antarctic ice sheet. *Journal of Glaciology* **45**, 273–285.

Manuel, N., Cornic, G., Aubert, S., Choler, P., Bligny, R. and Heber, U. (1999). Protection against photoinhibition in the alpine plant *Geum montanum*. *Oecologia* **119**, 149–158.

Maxwell, B. (1992). Arctic climate: Potential for change under global warming. *In* "Arctic Ecosystems in a Changing Climate: An Ecophysiological Perspective" (F. S. Chapin III, R. L. Jefferies, J. F. Reynolds, G. R. Shaver and J. Svoboda, eds), pp. 11–34. Academic Press, San Diego.

Nadelhoffer, K. J., Giblin, A. E., Shaver, G. R. and Laundre, J. A. (1992). Effects of temperature and substrate quality on element mineralization in six arctic soils. *Ecology* **72**, 242–253.

Neuner, G., Ambach, D. and Aichner, K. (1999). Impact of snow cover on photoinhibition and winter desiccation in evergreen *Rhododendron ferrugineum* leaves during subalpine winter. *Tree Physiology* **19**, 725–732.

Nobel, P. S. (1999). "Physicochemical and Environmental Plant Physiology". Academic Press, San Diego.

Oberbauer, S. F. and Oechel, W. C. (1989) Maximum CO_2 assimilation rates of vascular plants on an Alaskan arctic tundra slope. *Holarctic Ecology* **12**, 312–316.

Oberbauer, S. F., Cheng, W., Ostendorf, B., Sala, A., Gebauer, R., Gillespie, C. T., Virginia, R. A. and Tenhunen, J. D. (1996). Landscape patterns of carbon gas exchange in tundra ecosystems. *In* "Landscape Function and Disturbance in the Arctic" (J. F. Reynolds and J. D. Tenhunen, eds), pp. 223–257, Ecological Studies Series vol. 120. Springer Verlag, New York.

Oechel, W. C., Vourlitis, G. L., Hastings, S. J., Zulueta, R. C., Hinzman, L. and Kane, D. (2000). Acclimation of ecosystems CO_2 exchange in the Alaskan Arctic in response to decadal climate warming. *Nature* **406**, 978–981.

Pinuelas, J., Filella, I. and Gamon, J. A. (1995). Assessment of photosynthetic radiation-use efficiency with spectral reflectance. *New Phytologist* **131**, 291–296.

Reich, P. B., Walters, M. B. and Ellsworth, D. S. (1997) From tropics to tundra: global convergence in plant functioning. *Proceedings of the National Academy of Sciences*, **94**, 13 730–13 734.

Shaver, G. R., Billings, W. D., Chapin, F. S., Giblin, A. E., Nadelhoffer, K. J., Oechel, W. C. and Rastetter, E. B. (1992). Global change and the carbon balance of Arctic ecosystems. *Bioscience* **42**, 433–441.

Smith, W. K., Vogelmann, T. C., DeLucia, E. H., Bell, D. T. and Shepherd, K. A. (1997). Leaf form and photosynthesis: do leaf structure and orientation interact to regulate internal light and carbon dioxide? *BioScience* **47**, 785–793.

Starr, G. (2000). A multiscale assessment of physiological processes in arctic tundra plants under natural and simulated climate change scenarios. Ph.D. thesis, Florida International University, Miami, Florida.

Starr, G., Oberbauer, S. F. and Pop, E. W. (2000). Effects of lengthened growing season and soil warming on the phenology and physiology of *Polygonum bistorta*. *Global Change Biology* **6**, 357–369.

Streb, P., Shang, W., Feierabend, J. and Bligny, R. (1998). Divergent strategies of photoprotection in high-mountain plants. *Planta* **207**, 313–324.

Tieszen, L. L. (1981). Photosynthetic competence of the subnivean vegetation of an arctic tundra. *Arctic and Alpine Research* **6**, 253–256.

Walker, M. D., Walker, D. A. and Auerbach, N. A. (1994). Plant communities of a tussock tundra landscape in the Brooks Range foothills, Alaska. *Journal of Vegetation Science* **6**, 843–866.

Wookey, P. A., Welker, J. M., Parsons, A. N., Press, M. C., Callaghan, T. V. and Lee, J. A. (1994). Differential growth, allocation and photosynthetic responses of *Polygonum viviparum* to simulated environmental changes at a high arctic polar semi-desert. *Oikos* **70**, 131–139.

Anthocyanins in Autumn Leaf Senescence

DAVID W. LEE[1,2]

[1]*Department of Biological Sciences, Florida International University, Miami, FL 33199, USA*
[2]*Fairchild Tropical Garden, Miami, FL 33156, USA*

ABSTRACT

Anthocyanins are synthesized during leaf senescence in certain plants across virtually all biomes, but are most spectacular in the autumn foliage of temperate deciduous forests. The patterns of color production in senescing foliage depend at least partly upon species composition and their phenology. Both ecological and physiological explanations have been raised to explain why plants produce this pigment just before leaf fall. Physiological explanations, as photoprotection, predict that cyanic leaves would be better able to resorb nitrogen during the process of chlorophyll degradation. Ecological explanations predict better dispersal of propagules advertised by association with the brilliantly colored leaves (plausible for only a minority of species), or warning against egg-laying activity of herbivorous insects, as aphids. These hypotheses make predictions that we now can test, to help us understand this old mystery – and majestic phenomenon.

Advances in Botanical Research Vol. 37
incorporating Advances in Plant Pathology
ISBN 0-12-005937-1

I. INTRODUCTION

This volume explores the wide variety of circumstances under which anthocyanins are produced in vegetative organs. Yet, the appearance of red coloration during autumn leaf senescence, due to anthocyanin accumulation, is the most striking and widely appreciated of all of these phenomena. These autumn displays vary in intensity in different parts of the world, but are widespread in temperate deciduous forests. The spectacular shows of autumn color in forests of northeastern North America are major sources of revenue from tourism. Indeed, many readers will be attracted to this book because of past experiences of collecting colored autumn leaves during childhood, and living in settings where the forests change in color every year. Despite the high concentration of universities and other academic centers near such forests, surprisingly little research has been conducted on the phenomenon, and our base of scientific knowledge is rather meager. In this chapter I review this body of research, emphasizing recently published work (and some in which I have been involved), and will suggest directions for future research.

A. IMPORTANCE

Red and yellow foliage is produced during leaf senescence by different species in all temperate deciduous forests. No comparison of the extent and systematic distribution of color production has been completed, yet it is evident that the degree of color production is highly variable. The production of autumn coloration is perhaps most dramatic in the mixed deciduous forests of northeastern United States, particularly in New England. There, the red foliage is so pronounced that it is easily detectable by satellite remote sensing (Boyer *et al.*, 1988). However, the show of autumn coloration in the forests of the Great Smoky Mountains may be equally spectacular (Palevitz, 2001). Those in Europe and temperate Asia may be less pronounced, at least for the presence of brilliantly red-senescing trees and shrubs.

The autumn display of vermillion is an economic engine, drawing millions of visitors to these forests during the autumn, worth hundreds of millions of dollars in tourist revenues (Hendry, 1988). As a consequence, predictions of the timing and intensity of color production are of economic value. Local residents predict the intensity of color production based on the frequency of rainfall in the late summer months, as well as the frequency of cold and sunny days early in the autumn (John O'Keefe, personal communication).

In virtually all cases, the production of red foliage is due to the accumulation of anthocyanins in leaf tissues. The yellow colors are due to the persistence of xanthophyll pigments, made noticeable through the degradation of chlorophyll.

B. HISTORY

Naturalists and scientists have commented on autumn coloration for centuries, some of this documented by Wheldale's (1916) early review, along with the poetic descriptions of Kerner von Marilaum (1897). Early research on red coloration in New England forests was completed by Smith (1901), and Gertz (1906) included microscopic observations of senescent leaves in his extensive European survey of anthocyanin distribution in vegetative organs. Observations of this color production helped feed the speculations of late 19th century scientists arguing about the various possible functions of anthocyanins (see Lee and Gould in this volume). The small body of research published during the past 50 years will be discussed under specific headings in this chapter.

C. ANTHOCYANINS AND OTHER PIGMENTS

As a rule, orange–red coloration during senescence is caused by the synthesis of anthocyanins. There are a few cases of red coloration caused by xanthophylls (Ida *et al.*, 1995), but these are rare indeed. Additional research may reveal the production of even more unusual pigments during leaf senescence (see Lee in this volume), but this is also likely to be extremely unusual. Betacyanins are responsible for red coloration in leaves of species within the Caryophyllales, but this is apparently restricted to the formative stages in leaf ontogeny, or to foliage under physiological stress (Lee and Collins, 2001). Moreover, the woody species of Caryophyllales are almost entirely tropical evergreens. Given the higher metabolic cost of the nitrogenous betacyanins, their production during senescence should be less likely than anthocyanins.

The diversity of anthocyanin pigments in flowers is quite high; more than 250 different structures have been identified (Harborne and Grayer,

1988). Anthocyanin molecules vary in the structure of the pigment aglycone (anthocyanidin), as well as in the types of sugars and their positions of attachment on the anthocyanidin. In contrast, the structures of anthocyanins in senescing leaves are remarkably conserved; almost all are cyanidin based. Early work reported anthocyanins as predominantly cyanidin-3-glucosides (Ishakura, 1972). Given the application of the more sensitive and rapid contemporary techniques of high pressure liquid chromatography (HPLC), it is likely that other sugar moieties will be found in these leaf pigments with time (Ji *et al.*, 1992). Still, the low diversity of types of anthocyanins in leaves, compared to flowers, is striking.

D. UNMASKED OR SYNTHESIZED?

There is a common misperception about the origin of anthocyanins during leaf senescence. Textbooks frequently state that anthocyanins are permanently present during the leaf life span, their color production being unmasked by the breakdown of chlorophylls during senescence (Moore *et al.*, 1998; Uno *et al.*, 2001). The alleged accumulation of these pigments (Matile, 2000) is consistent with the classical argument of the accumulation in vacuoles of those by-products of metabolism with no physiological function other than carbohydrate overflow (Fraenkel, 1959; Luckner; 1984).

The unmasking hypothesis is scientifically unsound, however. Sanger (1971) had shown that anthocyanins are synthesized *de novo* in leaves during senescence. Lee *et al.* (2002) found no anatomical evidence of anthocyanins in mature leaves that accumulated these pigments in senescence (the 18 species studied in detail and referred to in other parts of this chapter are listed in Table I). Moreover, color production by anthocyanins in the cell vacuoles of mature leaves would modify the entire leaf color, as it does for certain tree cultivars, as the copper beech (*Fagus sylvatica* var. *atropunicea*) or purple varieties of the Japanese maple (*Acer palmatum* var. *atropurporeum*). So, these pigments would be visible in mature leaves. Extensive research in photobiology and in molecular developmental genetics (see Lee and Gould this volume), where anthocyanin induction and synthesis is a model system of research, has shown that anthocyanin synthesis is inducible in mature leaves by a variety of environmental and anthropogenic factors, (Chalker-Scott, 1999; Chalker-Scott in this volume). Clearly, anthocyanin synthesis during the senescence of red-foliage plants contrasts strongly with the production of coloration in yellow-foliage plants, which result from the unmasking of xanthophyll pigments during chlorophyll degradation.

TABLE I

Leaf senescence at the Harvard Forest, in Central Massachusetts USA (Lee et al., 2002). Eighteen species (with families) are listed, the species analysed for pigmentation and mentioned in the text; 16 species were monitored for phenological changes at weekly intervals 1991–1998 (except for 1992)

Species	Family	Mean time (± SD) to 50% leaf fall (Julian day)[a]	Anthocyanin distribution[b]				
			EP	PP	SM	UE	TS
Anthocyanic							
Acer rubrum L.	Aceraceae	280 ± 3	1	4	1	1	0
Acer saccharum H. Marsh	Aceraceae	293 ± 2	0	3	0	0	0
Cornus alternifolia L. Fil.	Cornaceae	287 ± 3	0	3	0	0	0
Fraxinus americana L.	Oleaceae	285 ± 4	0	4	0	0	0
Prunus serotina J. F. Ehrh.	Rosaceae	290 ± 3	1	0	0	0	0
Quercus rubra L.	Fagaceae	299 ± 4	0	3	1	0	0
Vaccinium corymbosum L.	Ericaceae	296 ± 4	0	4	0	0	0
Viburnum alnifolium H. Marsh	Caprifoliaceae	288 ± 6	0	4	0	0	0
Viburnum cassinoides L.	Caprifoliaceae	298 ± 7	0	4	1	0	0
Non-anthocyanic							
Acer pensylvanicum L.	Aceraceae	288 ± 5					
Betula alleghaniensis Britton	Betulaceae	281 ± 4					
Betula populifolia H. Marsh	Betulaceae	284 ± 4					
Castanea dentate (H. Marsh) Borkh.	Fagaceae	293 ± 4					
Fagus grandifolia J. F. Ehrh.	Fagaceae	306 ± 3					
Hamamelis virginiana L.	Hamamelidaceae	287 ± 4					
Ilex verticillata (L.) A. Gray	Aquifoliaceae	304 ± 4					
Populus grandidentata Michx.	Salicaceae	–					
Populus tremuloides *Michx.*	Salicaceae	–					
Overall mean		290 ± 6					

[a] Leaf drop is given as time in Julian days (consecutive days from the first of the year) to achieve 50% leaf fall. Figures are a mean of three lagged trees for each species.

[b] Scale of intensity of 1–4, where 1 = a few cells lightly pigmented and 4 = all cells densely pigmented. EP = adaxial epidermis; PP = palisade parenchyma; SM = spongy mesophyll; UE = abaxial epidermis; TS = trichomes/scales.

II. WHAT INFLUENCES PIGMENT DISTRIBUTION

Anthocyanin synthesis during leaf senescence is highly influenced by environmental factors. Concentrations can vary across different parts of a leaf, particularly in certain species such as *Acer rubrum* in eastern forests of the United States. In all species the concentration of anthocyanins varies among leaves from different parts of the tree or shrub crown, and is reduced by shading (Wheldale, 1916; Lee *et al.*, 2002). The shading of one part of a leaf by another suppresses anthocyanin production. Lee *et al.* (2002) also showed that alteration in red:far-red wavelengths (that could change phytochrome equilibria) had no perceptible effect on anthocyanin production.

Individual trees vary in production of anthocyanins during senescence. Such differences could be due to physical effects at microsites and/or genetic variation. For instance, some populations of aspen (*Populus tremuloides*) produce red leaves (Chang *et al.*, 1989) while most turn yellow during senescence.

Anthocyanin synthesis during the autumn is associated with leaf senescence and chlorophyll degradation. Both the overall process of leaf senescence (Smart, 1994; Quirino *et al.*, 2000) and chlorophyll degradation (Matile *et al.*, 1999; Thomas *et al.*, 2001) are highly coordinated and regulated processes. Given the association of anthocyanin synthesis with leaf senescence, chlorophyll concentration is a good physiological marker of production among different species (Wolf, 1956; Sanger, 1971; Koike, 1990; Collier and Thibodeau, 1995). The initial measurements of chlorophyll contents of 18 species at Harvard Forest (species listed in Table I), late in the growing season, indicate that considerable chlorophyll had already been lost by some of the species before anthocyanins were synthesised. Generally, anthocyanins began to appear in the leaves with about half of the initial chlorophyll (~20 $\mu g\ cm^{-2}$) remaining, but there is considerable variation among leaves, individuals and species.

A. PHENOLOGY

The timing of anthocyanin production during senescence varies among different species, as documented among nine species by Lee *et al.* (2002; Table I). Some species began senescence earlier than others, the earliest being *Acer rubrum* and the latest being *Fagus grandifolia* (whose leaves tend to be retained by the tree). What is most interesting in this phenology, from weekly observations by John O'Keefe over a 10-year period, is the strong consistency in the timing and order of senescence among these species from year to year. The intensity of color production, which is a combination of the number of leaves retained on the tree at peak coloration as well as the intensity of anthocyanin synthesis for the red leaves, varied from year to year. There is much speculation by residents living in the region as to what factors control the intensity of color. Adequate rainfall (but not too much) late in the growing season appears to be important, as do clear sunny days and cool nights in the period immediately preceding the induction of anthocyanin synthesis (Kozlowski and Pallardy, 1997).

B. SYSTEMATIC VARIATION

Some species with senescing red foliage will be found in virtually any forest. A small percentage of red-senescing trees have been documented in both tropical evergreen and deciduous forests (Lee and Collins, 2001; Lee in this volume). In forests of northeastern North America, this percentage may be much higher. For instance, at the Harvard Forest in central Massachusetts 62 of 89 woody species accumulated anthocyanins during senescence (Lee *et al.*, 2002; Plate 6). In some cases these accumulations resulted in spectacular color displays, but in others the combination of anthocyanins and residual chlorophyll produced brown coloration that appeared much like dead tissue. Some evergreen species may accumulate anthocyanins during the fall and winter and lose them at the beginning of the next growing season (see Starr and Oberbauer in this volume for such changes in Arctic evergreen vegetation, and Gould *et al.* this volume for New Zealand temperate evergreen forests).

The leaves of the large majority of these species produced anthocyanins in the mesophyll, predominantly in the palisade layer. For instance, in *Quercus rubra*, anthocyanins accumulated primarily in the palisade, late in the chlorophyll breakdown process (Plate 7). Anthocyanins were easy to detect in the vacuoles of these cells. Certain families may be particularly important sources of species producing red autumnal coloration, certainly the Aceraceae, Nyssaceae, Caprifoliaceae and Ericaceae. Others, such as the Betulaceae, or even sub-families, as the Fagoideae (but not the Quercoideae) are notable for the production of yellow autumn foliage and an absence of anthocyanins. In tropical forests certain families are particularly conspicuous for red coloration during senescence: particularly the Combretaceae and Lythraceae (see Lee in this volume). The brilliant senescent red leaves of the Indian almond (*Terminalia catappa*) are conspicuous in landscapes throughout the tropics, and in coastal forests of Southeast Asia. Thus the systematic composition of such forests may influence the patterns of color production.

C. CLIMATE

Classical authors noted the correlation between red foliage and temperature. Observations of the high incidence of red colors in alpine vegetation in Europe fed the arguments about anthocyanin function in leaves during the latter part of the 19th century (Wheldale, 1916; Kerner von Marilaum, 1897). Although less-studied, both the deciduous and the evergreen species of Arctic tundra plant communities turn red in the autumn (Starr and Oberbauer in this volume). Experimental evidence for the correlation between pigmentation and latitude (which would be equivalent to drop in temperature) was obtained by Townsend (1977) who grew accessions of the red maple (*Acer rubrum*) from different latitudes in a common garden. He observed that those from higher latitudes produced more autumn coloration. Hoch *et al.* (2001) concluded from a literature survey that among genera of woody species from North America and Europe, those species in northern North America were the more likely to produce anthocyanins during senescence. They also concluded that the milder maritime climates of European forests did not select for pigment production. However, in a survey of leaf anthocyanins in *Acer* reported by Ji *et al.* (1992) 67 out of 91 species (out of a total for the genus of approximately 200) produced pigments during senescence. Anthocyanic species were found at the north and south extremes of the geographical range, even in transient coloration among evergreen species from Southeast Asia.

Climatological gradients may result in ecotypic differentation, and ultimately the evolution of species with these pigmentation differences. Perhaps some of the differences in autumn coloration of temperate deciduous forests may be due to phylogenetic constraints within taxa that are geographically distributed among these forests (Lee in this volume).

D. ECOLOGICAL CORRELATES

Red foliage occurs in all of the plant functional groups that make up temperate deciduous forests. These include trees, shrubs, and perennials, all of these of varying successional stages. However, red leaf color may be less common among annuals in such forests (personal observation). At the Harvard Forest, the majority of species produced anthocyanins during senescence, including species among all of the functional groups listed above (Lee *et al.*, 2002).

In their review of the physiological significance of anthocyanins during autumnal leaf senescence Hoch *et al.* (2001) argued that species acclimated to higher insolation, such as early successional species, should be less susceptible to photoinhibition and damage, and therefore less likely to produce anthocyanins as additional protection during autumn leaf senescence. There is considerable evidence that plants adapted to high insolation are less susceptible to photoinhibition, and employ other mechanisms, such as the xanthophyll cycle pigments, to protect against photodamage (reviewed by Hoch *et al.*, 2001), also well-documented in high-altitude species (Streb *et al.*, 1998). Koike (1990) found that 16 early successional tree species in temperate deciduous forests in Japan began leaf senescence from the inside of their crowns towards the outside, and did not produce anthocyanins. Conversely, 14 late successional species began leaf senescence from the outside of their crowns and developed anthocyanins.

In their comparison of 18 species at Harvard Forest, Lee *et al.* (2002; Table I) did not observe a relationship between successional status and anthocyanin production. Also, some of the more shade tolerant species (theoretically the more susceptible to photoinhibition in high sunlight), such as *Acer pensylvanicum* and *Hamamelis virginiana*, did not produce anthocyanins during senescence. Another early successional species, *Prunus serotina*, was very weakly anthocyanic during senescence, compared to other species. However, it is important to remember that the New England forests, most famous for their spectacular autumn coloration, are not mature forests, since most of the region was converted to farmland in the 18th and 19th centuries.

At Harvard Forest, the nine red-senescing (and anthocyanic) species differed from the nine yellow-senescing species in several ways. Senescing leaves of the anthocyanic species had:

1. lower chlorophyll a/b ratios; and
2. greater mass/area.

Such differences between the two pigmentation patterns cannot be explained in terms of the successional status of the species.

III. THE FUNCTION OF ANTHOCYANINS DURING LEAF SENESCENCE

The chapters in this volume review past hypotheses on the possible physiological and other adaptive functions of anthocyanins in leaves. Authors have focused on five potential mechanisms that are objects of current research. These are:

1. photoprotection against photoinhibition and photodamage;
2. protection against UV-damage;
3. anti-oxidation;
4. osmotic regulation; and
5. biological interactions.

All of these potential explanations must confront a particular challenge in the case of autumn leaves: protection by anthocyanins appears an expensive option for leaves at the end of their life span, when photosynthetic capacity is low. Although anthocyanins have high extinction coefficients and can alter leaf optical properties at low concentrations (and are also extremely effective antioxidant molecules), they are the terminus of an elaborate biosynthetic pathway and their metabolic costs are relatively high. Thus, it seems likely that the benefit(s) to plants would extend beyond a few weeks at the end of the growing season.

A. PHYSIOLOGICAL EXPLANATIONS

The highly coordinated processes of leaf senescence (Smart, 1994; Quirino *et al.*, 2000) and chlorophyll degradation (Matile *et al.*, 1999; Thomas, 2001) appear to increase the resorption of nutrients, particularly nitrogen, back into woody plant tissues. Since almost all of leaf nitrogen is associated with photosynthesis, and much of that with chlorophyll in membrane-bound complexes, chlorophyll degradation is particularly important. Loss of nutrients from leaf fall varies among different plants and ecosystems (Aerts, 1996; Killingbeck, 1996). Recovery can be assessed in terms of its efficiency (or resorption efficiency of Killingbeck) but only if there is an accurate assessment of the maximum nitrogen (N) content during the leaf life span. Since the processes of senescence may begin during the growing season, multiple measurements may be necessary to determine this maximum. Then the efficiency can be assessed as:

$$\frac{(\text{N in leaves at maturity} - \text{N in falling leaves})}{\text{N in leaves at maturity}}$$

Resorption efficiencies averaged 50.3% across a large sample of species (Aerts, 1996). Because of the difficulty determining maximum N concentrations, Killingbeck (1996) has advocated the use of the term proficiency, which is the nitrogen content of the falling leaves. Values of N resorption proficiency in deciduous woody species varied from 0.46 to 1.90%.

Hoch *et al.* (2001), in an extensive survey of the literature, argued that anthocyanin accumulation during leaf senescence could increase N proficiency of deciduous plants by one or more protective mechanisms. A similar argument was raised independently (Feild *et al.*, 2001) in an experimental study of photoprotection during leaf senescence. Anthocyanins, through photoprotection by absorbing in the blue-green wavelengths could reduce photoinhibition and photodamage in senescing leaves (Smillie and Hetherington, 1999). The antioxidant properties of anthocyanins could also scavage reactive oxygen species (ROS) produced by a damaged photosynthetic apparatus (see Gould *et al.* this volume). This direct protection would facilitate a more efficient removal of nitrogen from mesophyll cells (where anthocyanins accumulate) back into woody stems. Movement of nitrogen requires adequate storage of carbohydrates for energy production and a permissable osmotic environment. Thus, although continued photosynthesis may not be critical for the overall carbon balance of the plant, it may provide a source of energy to allow the export of nitrogenous compounds to occur (Collier and Thibodeau, 1995; Solecka *et al.*, 1999; Merzlyak and Chivkunova, 2000). Indirect evidence supporting this hypothesis comes from a variety of sources, reviewed by Hoch *et al.* (2001). The hypothesis predicts a correlation of anthocyanin synthesis with chlorophyll decline. It also predicts that additional protection would be important where the potential of photodamage is increased. The high irradiances and low temperatures of autumn days, with high photosystem activity and the curtailing of dark CO_2 fixation by low temperatures, would increase potential for damage (Huner *et al.*, 1998).

Lee *et al.* (2002) tested for persistence in photoprotection of anthocyanic leaves in trees of nine red and nine yellow-senescing trees (Fig. 1). Using chlorophyll concentration as an indicator of senescence in leaves sampled during the autumn, they found a log-linear relationships in reduction of Fv/Fm with chlorophyll reduction. However, the

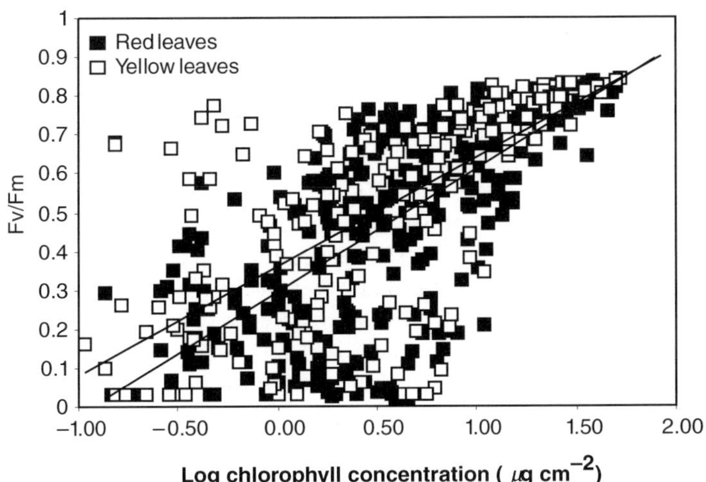

Fig. 1. Relationship between log of chlorophyll content (μg cm^{-2}) and Fv/Fm from leaves collected before dawn of nine red-senescing and nine yellow-senescing species from Harvard Forest, as listed in Table I (Lee *et al.*, 2002). A lower value for Fv/Fm indicates a reduction in quantum efficiency, most likely the consequence of photoinhibition.

slope of reduction of the red-senescing species was not significantly different from that for the yellow senescing species. Since the leaves used for these measurements were collected just before dawn, the plants had had ample opportunity for recovery from photoinhibition due to exposure the previous day. Feild *et al.* (2001, and Timmins *et al.* in this volume) conducted a more direct experiment by estimating photoinhibition and recovery in senescing red and green leaves of red-osier dogwood (*Cornus stolonifera*). They observed a significant reduction in photoinhibition, and more rapid recovery, in red leaves. These results give direct evidence for protection against photoinhibition in these senescing deciduous leaves. Krol *et al.* (1995) had seen a similar protective function in evergreen needles of *Pinus banksiana* seedlings, and Post and Vesk (1992) had detected similar photoprotection in Antarctic bryophytes. Starr and Oberbauer (in this volume) give supporting evidence for the operation of this protective mechanism in evergreen tundra plants. Other physiological mechanisms may also be important (see Chalker-Scott this volume in particular), but there is no direct supporting evidence at present.

A critical test for the operation of one or more of these protective mechanisms for the long-term survival of red-senescing plants is obviously to look at patterns in nitrogen resorption. Protection by anthocyanins should result in reduced nitrogen contents of falling leaves, and therefore more resorption into parent tissues. Obtaining such evidence is complicated by the effect of shading on anthocyanin production and, at the same time, reduced risk of photoinhibition. Using leaves that vary in anthocyanin production within a crown may not reveal any differences in nutrient resorption; Feild *et al.* (2001) failed to detect any differences in nitrogen resorption in senescent leaves of *C. stolonifera*.

Using a broader comparative approach, Lee *et al.* (2002) measured nitrogen in falling leaves of individuals of nine red- and nine yellow-senescing woody species at Harvard Forest (Table I for species compared). The nitrogen contents of senescent red-leaved species overall were reduced relative to those of the yellow, but the difference was not statistically significant ($P = 0.084$ from pairwise comparison by t test). However, when leaf samples of individuals were compared (Table II), leaf nitrogen (as a percentage of dry tissue mass) in senescent leaves was significantly and negatively correlated with anthocyanin concentration. In these comparisons, anthocyanins in senescent leaves were also correlated with mass/area of senescent leaves, as well as negatively correlated with nitrogen content in mature leaves of the same plants. However, anthocyanin content was not correlated with nitrogen content in leaves per unit area. Obviously, nutrient predictions of the protection hypothesesis need more extensive testing.

TABLE II
Pearson product correlations of leaf anthocyanin contents and other leaf characters in pooled leaf samples of three individuals from each of the 18 species at Harvard Forest, listed in Table I (Lee et al., 2002). All measurements were from senescent leaves, except for % nitrogen in mature leaf mass (%NM)

	Leaf characters[a]					
	ANT	CAS	MAS	%NS	%NM	NAS
ANT	–	0.004[f]	**0.362**	**–0.334**	**–0.361**	–0.052
Significance [b]		ns	**	*	**	ns
CAS		–	**0.404**	0.083	**0.308**	**0.353**
Significance			**	ns	**	*
MAS			–	0.067	0.105	**0.741**
Significance				ns	ns	**
%NS				–	**0.598**	**0.691**
Significance					**	**
%NM					–	**0.495**
Significance						**
NAS						–

[a] ANT = Anthocyanin concentration (μg cm $^{-2}$); CAS = chlorophyll/area (μg cm $^{-1}$); MAS = mass/area (mg cm $^{-2}$); %NS nitrogen as a % of senescent (dry) leaf mass; %NM = %nitrogen in mature leaf mass; NAS = nitrogen/area (μg cm $^{-2}$)
[b] Values in bold are significant, with * = < 0.05, ** = <0.005 and ns = not significant.

B. ECOLOGICAL EXPLANATIONS

Because of their strong visual signal (at least for animals with some color vision) it is natural to think of anthocyanic coloration as having some ecological significance. There has been some research and speculation on the ecological significance of anthocyanins of developing leaves in reducing herbivory (see review by Lee and Gould this volume), but avoiding herbivory on leaves that are about to abscise makes little sense. Nevertheless, there are two hypotheses invoking ecological interactions which would benefit the plant in subsequent growing seasons.

1. The Fruit Flag Hypothesis

Stiles (1982) proposed that brilliant leaf color in the autumn could attract birds to ripe fruits associated with the color, promoting disperal of propagules and enhancing the colonization success of the species. Birds are particularly attracted to red colors for flower visitation and pollination, as well as frugivory and seed dispersal. Such a mechanism requires that fruits be present on the plant at the end of the growing season when the leaves become visible. Obvious candidates for such a mechanism of attraction and dispersal are the sumacs (*Rhus* spp.) with their long infructescences at the ends of branches, associated with the subtending and brilliantly colored leaves. Several such species of *Rhus* were observed at Harvard Forest, all with high concentrations of anthocyanins in the palisade layer of their leaves. The most brilliantly colored autumn species there was a shrub, *Euonymus atropurpureus*, an invasive exotic common in the understory at forest fringes. In this species anthocyanin is produced at high concentrations in the adaxial epidermis predominantly when chlorophyll has been removed from the leaf mesophyll, producing the intense red color. The fruit flag hypothesis is certainly viable for plants that produce animal-dispersed propagules at the end of the growing season, but it can only account for pigment production in a small minority of trees in temperate deciduous forests. At Harvard Forest, most species produce fruits that ripen before autumn senescence begins, or their fruits are dispersed by wind or other mechanisms.

2. The Defensive Signal Hypothesis

If red leaf coloration were associated with lack of palatability of leaves or an increase in defensive compounds, such color could warn away potential herbivores. Coley and Barone (1996) summarized research that demonstrated such a defensive role in the flushing leaves of tropical rainforest plants. For those species, the putative warning signals had the obvious advantage of protecting expanding leaves at the beginning of their life spans. However, such a warning would not make sense for the protection of dying leaves. Hamilton and Brown (2001) adapted this hypothesis to the protection of deciduous-temperate tree species against the egg-laying and later herbivorous activity of aphids. They hypothesized that bright autumn coloration, associated with an increase in defensive chemistry, among different species and between populations or individuals within a single species, could deter late season visits by aphids. Such insects oviposit in the autumn, and the eggs hatch the following spring and increase herbivory pressure in the following year. Their test of this hypothesis was an analysis of published research, primarily a list of 262 tree species obtained from field guides which described the types and degrees of autumn coloration, and an exhaustive global compendium of aphids attacking trees (Blackman and Eastop, 1994). In their statistical regressions of autumn color against aphid diversity, they were able to account for potential phylogenetic influence (see Lee this volume). A model for these types of interactions was published by Archetti (2000) and was shown to be plausible.

Hamilton and Brown (2001) found a strong correlation between the degree of yellowness and herbivory by aphids (either all aphids or single specialist species), and a weaker relationship between redness and single specialist species. When they looked for regressions against autumn leaf color and attack by specialist aphids, and controlled for other variables, they found stronger regressions, but more so for yellow than for red color. They concluded that autumn color functions as a signal of defensive commitment against autumn colonization and future herbivory by aphids. In support of this hypothesis, there is good evidence that aphids discriminate among individuals within tree species, as shown by Furata (1990) for *Acer palmatum*.

The defensive signal hypothesis for anthocyanins in senescing leaves is deficient in several areas. First, the strongest signal against egg-laying is bright yellow coloration, the region of the spectrum where most aphids have sensitive color vision (Hamilton and Brown, 2001). Yellow color is the result of the differential rates of breakdown of chlorophylls and xanthophylls in chloroplasts. The rate of xanthophyll loss is the same in red- and yellow-senescing species at the Harvard Forest (Lee *et al.*, 2002; Fig. 2). Retention of xanthophylls in detached leaves does represent a carbon loss to these trees, but it is metabolically less expensive than the nitrogenous chlorophylls. Given the strong signal provided by yellow leaves, it seems illogical that anthocyanins would confer an additional advantage, unless certain specialist aphids have evolved color sensitivity in the red end of the spectrum. Finally, from a plant's standpoint, there is no evidence of a relationship between autumn color and defensive chemistry in leaves of any shrub or tree, let alone between autumn color and defensive compounds (such as tannins) in the following year.

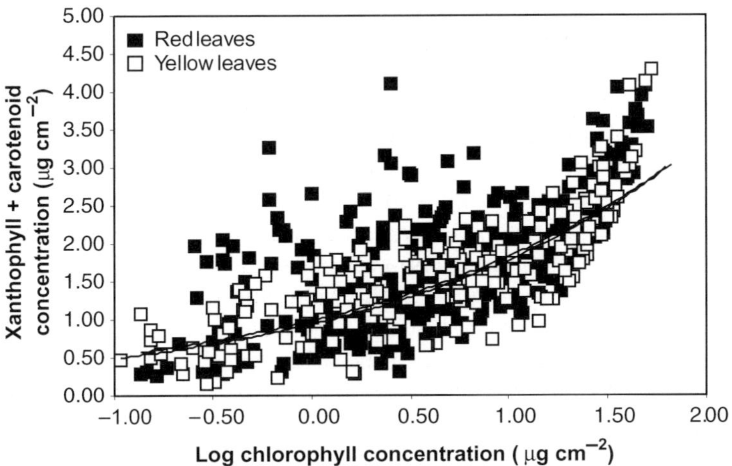

Fig. 2. Relationship between log of chlorophyll content (μg cm^{-2}) and total xanthophyll concentration (μg cm^{-2}) for leaves collected from the nine red-senescing and nine yellow-senescing species from Harvard Forest, as listed in Table I (Lee *et al.*, 2002). These plots indicate that the two species groups lose xanthophylls during senescence in an identical manner.

IV. FUTURE RESEARCH

We now have some good working hypotheses on how anthocyanins might function during leaf senescence, and these make predictions that can be tested by manipulating plants and measuring them in natural settings. We still need to learn more about physical factors that promote the production of anthocyanins, as well as the patterns (and their persistence) in anthocyanin production in populations of trees in forests. For instance, we need more empirical data on the intensity of anthocyanin coloration during autumn, and its relationship to climate and weather patterns during the year.

We need a plant that can be studied as a research system. The plant must grow quickly, be propagated easily, and be amenable to growth under greenhouse/growth chamber conditions. For such a plant we need to select and characterize mutants for deficiencies in anthocyanin production, so that we can compare their physiology under different stresses, such as cold and sunny conditions. Such a system would help us distinguish among the various physiological explanations that have been proposed for anthocyanin function. Ultimately, such a plant could be genetically engineered not to produce anthocyanins, then compared with normal wild type plants.

We need to study the nutrient dynamics, particularly in nitrogen flux, for species in forests with autumn coloration, to examine more critically the possibility of reduced nitrogen retention among species that produce anthocyanins during senescence. Again, comparisons among individuals within a species that vary in the production of anthocyanins would be a better test of the possible protective role of anthocyanins. Comparative approaches, where many individuals among numerous species with varying color production during senescence, would be a useful counterpoint to the more reductionist approaches.

We also need to pursue questions raised by the research on biological interactions. We need to examine carefully the color vision of aphid species that seem to specifically feed on anthocyanic trees. We also need to assess the relationship between anthocyanin production and secondary compound production (such as tannins) in natural populations and among different tree species.

It would also be interesting to look at the production of anthocyanins during leaf senescence in annual plants. Is such color production common? If so, what advantage would it be for a plant about to perish? Could such coloration contribute to higher nitrogen content (and survival) in seeds?

Finally, if anthocyanins are found to play a definitive role in the protection of plants during autumn senescence, then what alternative adaptive strategies have been selected by plants that do not turn red during leaf senescence?

V. CONCLUSION

Although the history of interest in autumn coloration stretches back over the centuries, only recently has that interest resulted in serious scientific study of the phenomenon. We now have some working hypotheses about anthocyanin function and clear indications on where to turn our attention next. The future looks bright for a much better understanding of this phenomenon in the near future.

ACKNOWLEDGMENTS

My interest and research in autumn coloration came late, after extensive experience in tropical forests. A Charles Bullard Fellowship at Harvard Forest in 1998 made it possible for me develop a more intimate appreciation for the phenomenon, and enabled me to form collaborations and friendships with John O'Keefe, Missy Holbrook and Taylor Feild. I would also like to acknowledge the remarkable career of W. D. Hamilton, cut short by tragic illness in Africa, whose contribution towards biological explanations of autumn coloration has stimulated much debate and will contribute to future research on this subject.

REFERENCES

Aerts, R. (1996). Nutrient resorption from senescing leaves of perennials: are there general patterns? *Journal of Ecology* **84**, 597–608.

Archetti, M. (2000). The origin of autumn colours by coevolution. *Journal of Theoretical Biology* **205**, 625–630.

Blackman, R. L. and Eastop, V. F. (1994). "Aphids on the World's Trees". CABI, Wallingford, UK.

Boyer, M., Miller, J., Belanger, M., Hare, E. and Wu, J. (1988). Senescence and spectral reflectance in leaves of northern pin oak (*Quercus palustris* Muenchh.). *Remote Sensing of the Environment* **25**, 71–87.

Chalker-Scott, L. (1999). Environmental significance of anthocyanins in plant stress responses. *Photochemistry and Photobiology* **70**, 1–9.

Chang, K. G., Fechner, G. H. and Schroeder, H. A. (1989). Anthocyanins in autumn leaves of quaking aspen in Colorado. *Forest Science* **35**, 229–236.

Coley, P. D. and Barone, J. A. (1996). Herbivory and plant defense in tropical forests. *Annual Review of Ecology and Systematics* **27**, 305–335.

Collier, D. E. and Thibodeau, B. A. (1995). Changes in respiration and chemical content during autumnal senescence of *Populus tremuloides* and *Quercus rubra* leaves. *Tree Physiology* **15**, 759–764.

Feild, T. S., Lee, D. W. and Holbrook, N. M. (2001). Why leaves turn red in autumn. The role of anthocyanins in senescing leaves of red-osier dogwood. *Plant Physiology* **127**, 566–574.

Fraenkel, G. S. (1959). The raison d'être of secondary plant substances. *Science* **129**, 1466–1471.

Furata, K. (1990). Early budding of *Acer palmatum* caused by the shade; intraspecific heterogeneity of the host for the maple aphid. *Bulletin of the Tokyo University Forests* **82**, 137–145.

Gertz, O. (1906). "Studier öfver Anthocyan". Akademisk Afhandling, Lund.

Hamilton, W. D. and Brown, S. P. (2001). Autumn tree colours as a handicap signal. *Proceedings of the Royal Society of London B. Biological Science* **268**, 1489–1493.

Harborne, J. B. and Grayer, R. J. (1988). The anthocyanins. *In* "The Flavonoids, Advances in Research Since 1980" (J. B. Harborne, ed.), pp. 1–20. Academic Press, New York.

Hendry, G. A. F. (1988). Where does all the green go? *New Scientist* **1637**, 38–42.

Hoch, W. A., Zeldin, E. L. and McGown, B. H. (2001). Physiological significance of anthocyanins during autumnal leaf senescence. *Tree Physiology* **21**, 1–8.

Huner, N. P. A., Oquist, G. and Sarhan, F. (1998). Energy balance and acclimation to light and cold. *Trends in Plant Science* **3**, 224–230.

Ida, A., Masamoto, K., Maoka, T. and Sarhan, F. (1995). The leaves of the common box, *Buxus sempervirens*, become red as the level of a red carotenoid, anhydroescholtzxanthin, increases. *Journal of Plant Research* **108**, 369–376

Ishikura, N. (1972). Anthocyanins and other phenolics in autumn leaves. *Phytochemistry* **11**, 2555–2558.

Ji, S.-B., Yokoi, M., Saito, N. and Mao, L.-S. (1992). Distribution of anthocyanins in Aceraceae leaves. *Biochemical Systematics and Ecology* **20**, 771–781.

Kerner von Marilaum, A. (1897). "The Natural History of Plants" (translated by F. W. Oliver). Blackie, London.

Killingbeck, K. T. (1996). Nutrients in senesced leaves: keys to the search for potential resorption and resorption proficiency. *Ecology* **77**, 1716–1727.

Koike, T. (1990). Autumn coloring, photosynthetic performance and leaf development of deciduous broad-leaved trees in relation to forest succession. *Tree Physiology* **7**, 21–32.

Kozlowski, T. T. and Pallardy, S. D. (1997). "Physiology of Woody Plants". Academic Press, San Diego.

Krol, M., Gray, G. R., Hurry, V. M., Öquist, G., Malek, L. and Huner, N. P. A. (1995). Low-temperature stress and photoperiod affect an increased tolerance to photoinhibition in *Pinus banksiana* seedlings. *Canadian Journal of Botany* **73**, 1119–1127.

Lee, D. W. and Collins, T. M. (2001). Phylogenetic and ontogenetic influences on the distribution of anthocyanins and betacyanins in leaves of tropical plants. *International Journal of Plant Science* **162**, 1141–1153.

Lee, D. W., O' Keefe, J., Holbrook, N. M. and Feild, T. S. (2002). Pigment dynamics and autumn leaf senescence in a New England deciduous forest. Submitted for publication.

Luckner, M. (1984). "Secondary Metabolism in Microorganisms, Plants, and Animals". Springer-Verlag, Berlin.

Matile, P. (2000). Biochemistry of Indian summer: physiology of autumnal leaf coloration. *Experimental Gerontology* **35**, 145–158.

Matile, P., Hörtensteiner, S. and Thomas, H. (1999). Chlorophyll degradation. *Annual Review of Plant Physiology and Plant Molecular Biology* **50**, 67–95.

Merzlyak, M. N. and Chivkunova, O. B. (2000). Light-stress-induced pigment changes and evidence for anthocyanin photoprotection in apples. *Journal of Photochemistry and Photobiology* B **53**, 155–163.

Moore, R., Clark, W. D. and Vodopich, D. S. (1998). "Botany", 2nd edn. WCB/McGraw-Hill, Dubuque, IA.

Palevitz, B. A. (2001). Why leaves turn color in the fall. *The Scientist* **15**(24), 8.
Post, A. and Vesk, M. (1992). Photosynthesis pigments and chloroplast ultrastructure of an Antarctic liverwort from sun-exposed and shaded sites. *Canadian Journal of Botany* **70**, 2259–2264.
Quirino, B. F., Noh, Y.-S., Himelblau, E. and Amasino, R. M. (2000). Molecular aspects of leaf senescence. *Trends in Plant Science* **5**, 278–282.
Sanger, J. (1971). Quantitative investigations of leaf pigments from their inception in buds through autumn coloration to decomposition in falling leaves. *Ecology* **52**, 1075–1089.
Smart, B. S. (1994). Gene expression during leaf senescence. *New Phytologist* **126**, 419–448.
Smillie, R. M. and Hetherington, S. E. (1999). Photoabatement by anthocyanin shields photosynthetic systems from light stress. *Photosynthetica* **36**, 451–463.
Smith, F. G. (1901). On the distribution of red color in vegetative parts in the New England flora. *Botanical Gazette* **32**, 332–342.
Solecka, D., Boudet, A. and Kacperska, A. (1999). Phenylpropanoid and anthocyanin changes in low-temperature treated winter oilseed rape leaves. *Plant Physiology and Biochemistry* **37**, 491–496.
Stiles, E. W. (1982). Fruit flags: two hypotheses. Fruit colors provide signals to dispersing birds. *American Naturalist* **120**, 500–509.
Streb, P., Shang, W., Feierabend, J. and Bligny, R. (1998). Divergent strategies of photoprotection in high-mountain plants. *Planta* **207**, 313–324.
Thomas, H., Ougham, H. and Hörtensteiner, S. (2001). Recent advances in the cell biology of chlorophyll catabolism. *Advances in Botanical Research* **35**, 1–52.
Townsend, A. M. (1977). Characteristics of red maple progenies from different geographic areas. *Journal of the American Society of Horticultural Science* **102**, 461–466.
Uno, G., Storey, R. and Moore, R. (2001). "Principles of Botany". McGraw-Hill, New York.
Wheldale, M. (1916). "The Anthocyanin Pigments of Plants". Cambridge University Press, Cambridge.
Wolf, F. T. (1956). Changes in chlorophylls a and b in autumn leaves. *American Journal of Botany* **43**, 714–718.

A Unified Explanation for Anthocyanins in Leaves?

KEVIN S. GOULD[1], SAM O. NEILL[1] and THOMAS C. VOGELMANN[2]

[1]School of Biological Sciences, University of Auckland, Private Bag 92019, Auckland, New Zealand
[2]Department of Botany and Agricultural Biochemistry, The University of Vermont, Burlington, VT 05405, USA

ABSTRACT

The leaves from many of New Zealand's native species are remarkably polymorphic for anthocyanin expression. Red coloration varies not only as a function of seasonal and developmental factors, but can also differ among individuals of a population, among leaves within a canopy, and even among tissues within a leaf. Moreover, the biosynthesis of anthocyanin in these leaves can be induced by a host of disparate environmental and biotic stimuli. Any unified explanation for the presence of anthocyanins in leaves must accommodate both the variability in pigmentation patterns over time and space, and the diverse range of triggers. Our data indicate that anthocyanins confer a phytoprotective role, rather than being the default end-product of a saturated flavonoid metabolism. Anthocyanins are primarily associated with chlorophyllous tissues, and significantly modify both the quantity and the quality of light incident on a chloroplast. Red leaves photosynthesise less than green leaves, but are also photoinhibited less and recover sooner following exposure to high light fluxes. Photoabatement also reduces the generation of free radicals and reactive oxygen species from photooxidation, photorespiration, and Mehler reaction activities. Anthocyanins inhibit Fenton hydroxyl radical generation by chelating to ferrous ions, and effectively scavenge superoxide and hydrogen peroxide generated by mechanical injury, sudden temperature changes, and exposures to high light. Anthocyanins are evidently versatile and highly effective phytoprotectants. However, there is probably no unified explanation for their presence in leaves. Common among the first land plants, anthocyanins have probably been hijacked over the course of evolution to perform an array of tasks.

Advances in Botanical Research Vol. 37
incorporating Advances in Plant Pathology
ISBN 0-12-005937-1

I. INTRODUCTION

Possible functions of red pigments in leaves have been the focus of scientific enquiry for more than a century, and the hypotheses are almost as diverse as they are numerous. Photoprotection, UV-B screening, enhanced light capture, desiccation tolerance, cold-hardiness, anti-fungal properties, and camouflage from herbivores – all have been ascribed to the presence of anthocyanins in leaves, and most are supported by solid bodies of data (reviewed by McClure, 1975; Chalker-Scott, 1999). However, none adequately accounts for all instances of anthocyanin accumulation across the plant kingdom, and contemporary scientists are far from arriving at a consensus as to the biological role of these pigments. Can there be a unified explanation for anthocyanins in leaves?

The quest for an all-embracing functional hypothesis is fraught with potential difficulties. In the first place, a unified explanation must accommodate the widely disparate environments in which red leaves are a significant feature. Cyanic leaf laminae are commonly encountered in arid environments as well as the humid understorey of tropical rainforests. Desiccation tolerance or UV-B protection might account for the presence of anthocyanins in succulent plants such as *Cotyledon orbiculata* (Barker *et al.*, 1997), but they do not explain why understorey ferns and herbs have red leaf undersurfaces in their dark, moist habitats (Lee *et al.*, 1979). Similarly, a leaf-warming role could benefit the Antarctic liverwort *Cephaloziella exiflora* (Post and Vesk, 1992), but this effect would surely be disadvantageous in the tropics for the exposed, rapidly expanding leaves of mango and cacao (Lee *et al.*, 1987).

A unified explanation for anthocyanin function would also need to account for the remarkable diversity in anthocyanin expression. Patterns of red pigmentation vary across species, among individuals of a population, and sometimes among leaves on a single plant (Plate 8 A–F).

Species from similar habitats can be pigmented on the upper, lower, or both lamina surfaces (Gould and Quinn, 1999). For many species, anthocyanins are restricted to only a fraction of the total leaf area, often at the leaf margin, midrib or petiole, or as irregular spots, patches or stripes against the green background. Natural polymorphism for anthocyanin expression is particularly common among the warm-temperate flora of New Zealand. *Quintinia serrata* (Escalloniaceae), an evergreen canopy tree from New Zealand's North Island, shows up to eight distinct patterns of leaf pigmentation – incremental steps in anthocyanin coverage of the lamina surface – and branches bearing mixtures of red and green leaves are common (Plate 8 A; Gould *et al.*, 2000). In species such as *Q. serrata*, the requirements for phytoprotection by anthocyanins are evidently highly localised. A unified hypothesis for anthocyanin function has both to account for the localised requirement for pigmentation, and also explain how green leaves, or green portions of otherwise red laminae, circumvent this requirement.

Many of the contemporary hypotheses for anthocyanin function require that the pigments reside at specific cellular locations within the leaf for optimal effectiveness. For example, anthocyanins must be held in the upper epidermis to screen UV-B, or in the mesophyll to protect chloroplasts from photoinhibition. Anthocyanins are most commonly found in the vacuoles of the palisade mesophyll and/or the spongy mesophyll cells, but they can also occur in the upper and lower epidermises, the hypodermis, idioblasts, trichomes, and/or the parenchyma associated with vascular bundles (Gould and Quinn, 1999; Lee and Collins, 2001). The histological distribution of anthocyanins is usually taxon specific, though there are exceptions. In the polymorphic species *Q. serrata*, for example, 14 different red-tissue combinations have been observed (Gould *et al.*, 2000). Variation in anthocyanin location extends to the sub-cellular level. Although the pigments are normally present as a solution inside the cell vacuole (Lee and Collins, 2001), they have also been observed as intensely pigmented intravacuolar vesicles (anthocyanoplasts) in more than 70 species (Peckett and Small, 1980), in associations with protein matrices in the cell vacuoles of petals and tubers (Nozue *et al.*, 1995; Markham *et al.*, 2000), and as components of cell walls in the leaves of bryophytes (Kunz *et al.*, 1994). Because of this cellular and sub-cellular variation, any location-specific hypothesis for anthocyanin function must be rejected as a contender for the universal explanation.

The timing of anthocyanin expression also differs markedly among species. Some plants always bear cyanic leaves, but many more show ontogenetic, and/or seasonal variation. Anthocyanin synthesis may be exclusive to the young, developing leaves, such as those of woody species from the tropical rainforest (Lee and Lowry, 1980; Woodall *et al.*, 1998), or else be upregulated during the terminal stages in leaf senescence, as in many deciduous trees of temperate regions (Hoch *et al.*, 2001; Feild *et al.*, 2001). A number of woody perennial taxa produce red leaves during the juvenile, but not the adult phase of the plant's ontogeny (Hackett, 1985, and this volume). Others show seasonal fluctuations in anthocyanin levels in response to environmental cues such as low temperatures and drought (Chalker-Scott, 1999). Interactions are commonplace: the New Zealand podocarp *Dacrydium cupressinum*, for example, produces red needles during the winter months, but only at the tips of young, flushing shoots (Plate 8 D), and only in the juvenile phase of the tree's ontogeny (Hinchliff, 2001). The regulation of anthocyanin production in such species is evidently highly complex.

Thus, diversity in both the timing and the expression of anthocyanin production presents a significant obstacle to our understanding of the potential functions of these pigments in leaves. Paradoxically, this natural variability also provides us with a powerful and versatile scientific tool to explore anthocyanin function. It facilitates comparisons of the performance of a red leaf versus a green leaf, a red shoot versus a green shoot, and at the microscopic level, a red cell versus a green cell, in response to prescribed environmental challenges. Knowledge of the functional consequences of heterogeneity in anthocyanin expression provides the key to understanding why some plants produce red leaves.

Amidst the apparent chaos of permutations in red pigmentation, there are at least three features common to most, if not all, natural populations of cyanic plants. (Here, we draw the distinction between the properties of natural cyanic populations, and those of ornamentals such as *Coleus*, which have been specifically bred for enhanced anthocyanin production. Potential commonalities between anthocyanin function in these and the naturally occurring red leaves are unknown). The anthocyanins

1. absorb visible radiation,
2. are effective scavengers of active oxygen species, and
3. are inducible by a wide range of environmental, biotic and/or anthropogenic stressors.

The experimental evidence for, and functional implications of each of these features are discussed below in detail.

We should state at the outset our default position – that anthocyanins do, indeed, confer an advantage to leaf function. It is possible, of course, that anthocyanins are simply the ergastic end-products of a saturated flavonoid metabolism, which are sequestered into the cell vacuole for storage. Anthocyanins are prominent to us because of their red colour, but it may be that they are less effective phytoprotectants than their colourless flavonoid precursors. However, our own data, and those of contemporary scientists world-wide, would indicate otherwise. The presence of anthocyanins can have a profound impact on the photophysiology of a leaf, at least in the species that have been studied so far. In the following sections we describe these effects, and develop our argument that a primary function of anthocyanins in leaves is to protect tissues from the effects of oxidative stress.

II. PHOTOPHYSIOLOGY OF CYANIC LEAVES

A. LIGHT ABSORPTION PROFILES

The anthocyanins, like all other flavonoids, absorb solar radiation. Acidified solutions of anthocyanin glycosides typically have two absorption maxima, one in the visible region between 465 and 550 nm, and a smaller one in the ultraviolet at about 275 nm (Harborne, 1967). Variations in hue – the reds, pinks, mauves and blues – are attributable to differences in the nature and degree of substitution of the anthocyanin molecules, the pH of their aqueous environment, possible conjugation to proteins, and co-pigmentation effects (Brouillard and Dangles, 1994; Markham *et al.*, 2000).

The property of anthocyanins to absorb radiation significantly modifies both the quality and the quantity of light that is absorbed by a leaf. Red leaves typically absorb more radiation in the 500–600 nm waveband than do their green counterparts (Fig. 1A), the magnitude of which is directly proportional to anthocyanin concentrations in the leaf (Neill and Gould, 1999). The effect of anthocyanins on leaf optics can be considerable; in *Q. serrata*, for example, red leaves absorb up to 17% more photosynthetically active radiation (PAR) than green leaves, but this figure differs markedly across other species (Eller *et al.*, 1981; Gausman, 1982; Burger and Edwards, 1996; Woodall *et al.*, 1998). The enhanced absorptance by red leaves is largely independent of the histological locations of cyanic cells in the lamina (Fig. 1A). Red leaves do not usually reflect more red light than do green leaves (Fig. 1B); human perception of red colouration is attributable to the subtraction of green and yellow wavelengths from the spectrum reflected from the lamina surface, rather than to any enhancement in red reflectance (Neill and Gould, 1999).

What is the fate of these absorbed quanta? Photoprotective hypotheses for anthocyanins in leaves implicitly assume that the light energy absorbed by the anthocyanins cannot be transferred to the chloroplasts. This seems a reasonable assumption; anthocyanins most frequently reside in the cell vacuole and are, therefore, physically separated from the photosynthetic machinery. Accordingly, anthocyanins function as simple light filters, and probably re-dissipate the absorbed energy as heat.

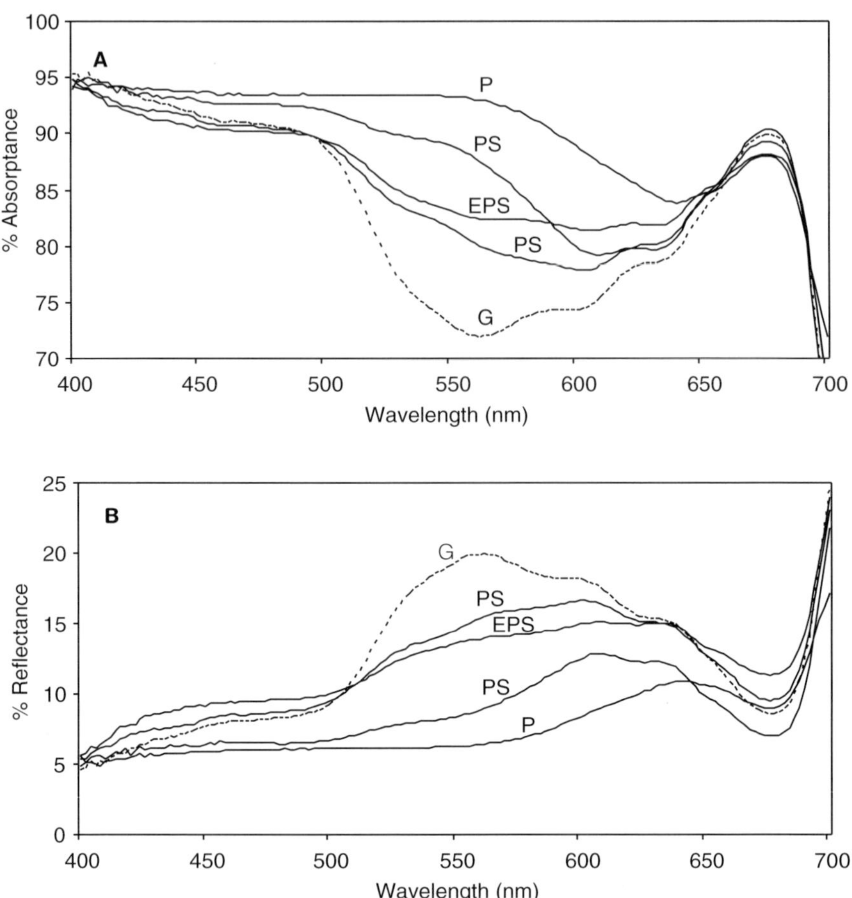

Fig. 1. Absorptance (A) and reflectance (B) spectra for one green (G) and four red leaves of *Quintinia serrata*. Anthocyanins were located in upper and lower epidermises (E), palisade mesophyll (P), and/or spongy mesophyll (S). Variation in optical properties among red leaves was attributable to differences in anthocyanin concentration, rather than histological location of the pigment. Modified from Neill and Gould (1999).

However, in the absence of cellular information we cannot exclude the possibility that some or all of the energy absorbed by anthocyanins is re-emitted as light that is useful for photosynthesis. Anthocyanins emit blue and green fluorescence when excited by ultraviolet radiation (Drabent *et al.*, 1999), and anthocyanin-rich extracts have been reported to enhance the Hill reaction activity of chloroplast suspensions from several species (Sharma and Banerji, 1981; Dhawale *et al.*, 1983). Moreover, photo-excited anthocyanins have a proven capacity to transfer electrons to semiconducting materials inside a solar cell (Cherepy *et al.*, 1997). The possibility of energy transfer from anthocyanins in biological systems is not inconceivable.

The issue of light utilisation in cyanic leaves was recently addressed using the methodology of Vogelmann and Han (2000). Transverse sections through red and green portions of *Q. serrata* leaf laminae were irradiated on their adaxial surfaces with monochromatic blue, red and green light, and the profiles of absorbed light determined from images of red chlorophyll fluorescence at the cut face (Gould *et al.*, 2002a). All laminae, both green and red, showed a descending gradient in light utilisation from the top to the bottom of the leaf, a feature that is common to dorsiventral leaves of many taxa (Smith *et al.*, 1997). However, the presence of anthocyanins greatly accentuated those gradients (Fig. 2). Anthocyanins in the mesophyll restricted absorption of green light, and to a lesser extent red light, to chloroplasts within the uppermost palisade mesophyll cells. When anthocyanins were also present in the upper epidermis, light absorption was further restricted to an extremely narrow band at the epidermis-palisade boundary. Thus, anthocyanins in the uppermost leaf tissues effectively deprive the spongy mesophyll of green light.

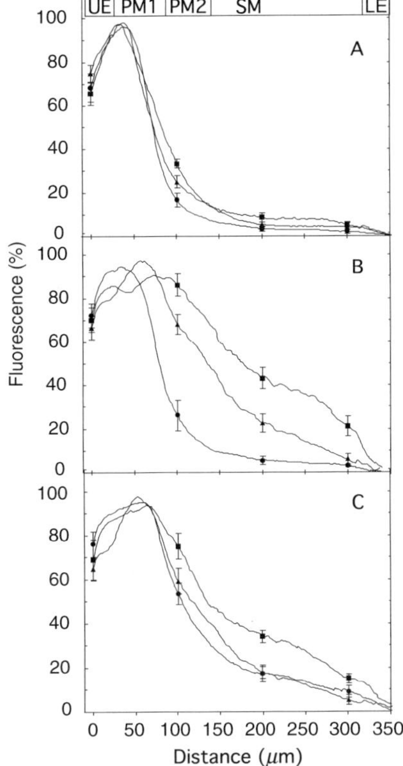

Fig. 2. Chlorophyll fluorescence profiles for transverse sections through leaves of *Q. serrata* irradiated with monochromatic blue (A), green (B) and red (C) light. Anthocyanins were absent (■), were present in all cell layers (●), or were distributed irregularly through the palisade and spongy mesophyll (▲). Relative positions of upper epidermis (UE), uppermost palisade mesophyll (PM1), second palisade mesophyll layer (PM2), spongy mesophyll (SM), and lower epidermis (LE) are indicated. Symbols identify the lines; actual data points are spaced 1.8 μm apart. Each line is the average of nine profiles on three leaves (●), or six profiles on two leaves (■,▲). Error bars show SE. From Gould *et al.* (2002a).

B. PHOTOSYNTHESIS

Green light is absorbed by chlorophyll less efficiently than red or blue light. Nonetheless, green light contributes significantly to the photosynthesis of green leaves, particularly in the lower mesophyll tissues (Sun *et al.*, 1998; Nishio, 2000). The screening of green light by a cyanic filter would reduce the quantum efficiency of photosynthesis with respect to the total light absorbed, and increase the threshold irradiance at which light-saturated photosynthesis is achieved. Prolonged exposure to the filtered light can also lead to permanent alterations in physiology as a result of chloroplasts developing under, or acclimating to the shaded conditions.

Many reports indicate that net photosynthesis is lower in red than in green leaves. For example, species of *Syzygium* which produce cyanic juvenile leaves have lower photosynthetic capacities than related species with green juvenile leaves (Dodd *et al.*, 1998), and the red, flushing leaves of *C. orbiculata* have lower dark-adapted efficiencies of photosystem II (PSII) than do the mature, green leaves (Barker *et al.*, 1997). Similarly, maximum quantum yields for photosynthetic oxygen evolution are significantly lower for red-leaf varieties of *Coleus* than for green-leaf varieties (Burger and Edwards, 1996), and light-saturated rates of carbon assimilation in the rapidly expanding leaves of *Brachystegia spiciformis* are inversely correlated to red pigment levels (Tuohy and Choinski, 1990; Choinski and Johnson, 1993). There are, however, notable exceptions. Among the understorey plants of tropical rainforests, red-leafed morphs have significantly higher light-saturated rates of carbon assimilation than the green morphs (Gould *et al.*, 1995). For those plants, the anthocyanins are usually located in cells subjacent to the chlorenchyma; the red leaves apparently benefit from the presence of red pigments without compromising their already limited capacities to capture light.

For *Q. serrata*, the screening of quanta by cyanic cells in the palisade mesophyll translates into a 23% reduction in CO_2 assimilation under light-saturating conditions, and a lower threshold irradiance for light-saturation, compared to those of green leaves (Fig. 3A). Differences between the photosynthesis of the red and green leaves are similar to those for shade and sun species, and for sun species acclimated to low- and high-light environments (Björkman, 1981). They are indicative of the long-term acclimation to light abatement by chloroplasts subjacent to a cyanic filter.

Shade-acclimation in red *Q. serrata* leaves is particularly evident among cells of the spongy mesophyll. When thick, transverse sections through red leaves were irradiated at their cut face by a narrow, linear beam of modulated red light (Han and Vogelmann, 1999), photoacoustic signals for oxygen evolution were restricted almost exclusively to the

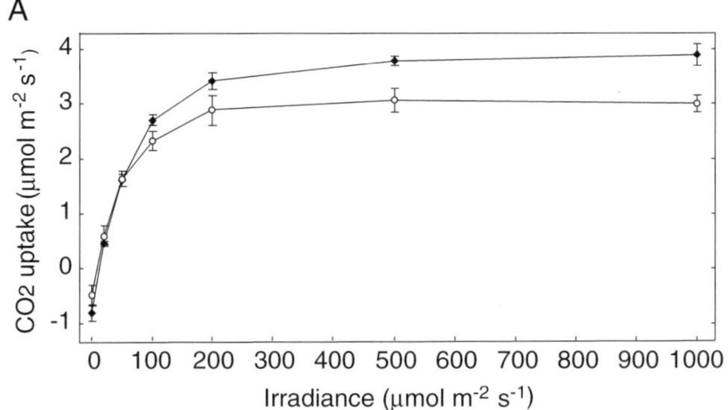

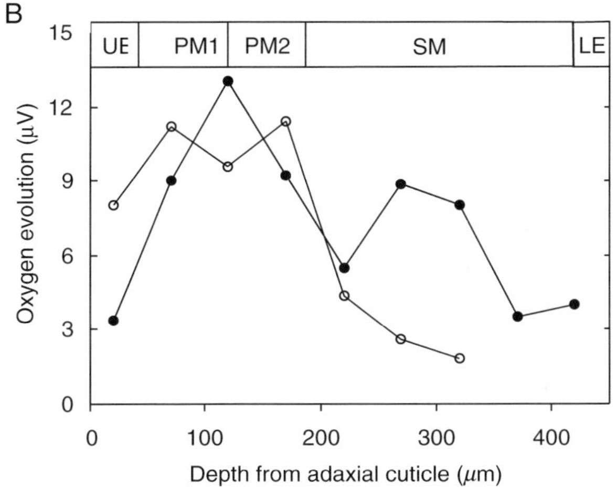

Fig. 3. Light-response curves for CO_2 assimilation (A) and profiles for oxygen evolution capacity (B) in green (●) and red (○) leaf portions of *Quintinia serrata*. Relative positions of upper epidermis (UE), uppermost palisade mesophyll (PM1), second palisade mesophyll layer (PM2), spongy mesophyll (SM), and lower epidermis (LE) are indicated. Error bars show SE. From Gould *et al.* (2002a).

cells of the palisade mesophyll (Fig. 3B). Oxygen evolution from spongy mesophyll cells was substantially lower than that from the palisade mesophyll of the same red leaves. Green leaves, by contrast, yielded photoacoustic signals for oxygen evolution from both the palisade and the spongy mesophyll (Fig. 3B). Anthocyanins not only reduce the photosynthetic performance of red versus green leaves, they also reduce the *potential* for cells in the spongy mesophyll cells to photosynthesise.

C. PHOTOINHIBITION

Clearly, reductions in the photosynthetic capabilities of red leaves would not *per se* benefit the carbon economy of a plant. However, by intercepting high-energy quanta that would otherwise be absorbed by chlorophyll b, the anthocyanins present a mechanism for the protection of the photosynthetic apparatus from photoinhibition under conditions of high irradiance. The presence of anthocyanins has been associated with reductions in the severity of dynamic photoinhibition in a number of species (Gould *et al.*, 1995; Krol *et al.*, 1995; Smillie and Hetherington, 1999; Feild *et al.*, 2001), although exceptions have been noted (Burger and Edwards, 1996).

Photoinhibitory responses of red and green leaves are often difficult to evaluate because of the confounding differences in leaf age between the two types (Dodd *et al.*, 1998). This problem was circumvented recently using the New Zealand herb *Elatostema rugosum* (Urticaceae). An inhabitant of damp, shaded areas, *E. rugosum* is exceptionally polymorphic for anthocyanin production in leaves (Plate 8 B). Leaves of red morphs are intensely cyanic when young, but become predominantly green after they have expanded fully. The green morphs, which frequent similar habitats, bear only green leaves at all stages in development (Smith, 2000). The polymorphism afforded comparisons of light responses in red and green leaves of similar ages. Photosynthesis of both red and green leaves is light-saturated with an irradiance of approximately 300 μmol m^{-2} s^{-1}, but under natural conditions the leaves may receive stray beams of light more than five times as intense. When dark-adapted *E. rugosum* plants were subjected to 30 minutes of white light at 1400 μmol m^{-2} s^{-1} and then returned to darkness, the quantum efficiencies (Fv/Fm values) of green and red leaves initially fell to 52% and 62% of their maximum values, respectively (Fig. 4). Red leaves recovered almost completely within 20 minutes after their return to darkness. The green leaves, by contrast, recovered to the same degree only 40 minutes after the treatment. The attenuation of light by anthocyanins thus has the potential both to reduce the severity of, and to hasten the recovery from dynamic photoinhibition. Photoprotection of this nature would be particularly useful for shade-adapted plants subjected to transient, high-intensity sunflecks (Gould *et al.*, 1995).

D. PHOTOOXIDATION

Photoabatement by anthocyanins in leaves also has the potential to prevent or reduce photooxidative damage. When chlorenchyma are exposed to light energy in excess of their requirements for carbon assimilation, they generate oxygen radicals and other reactive oxygen species

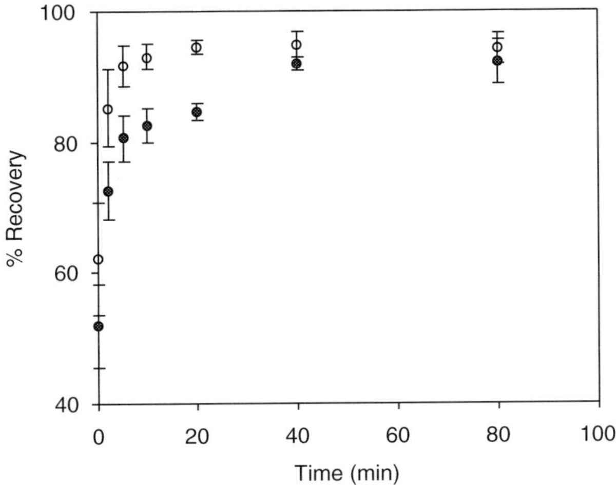

Fig. 4. Recovery of initial, dark-adapted Fv/Fm values in red (○) and green (●) leaves of *Elatostema rugosum* after 30 min irradiation with 1400 μmol m^{-2} s^{-1} white light. From Smith (2000).

(ROS) that can cause structural damage to DNA, lipid membranes, and proteins (reviewed by Foyer *et al.*, 1994; Alscher *et al.*, 1997; Polle, 1997). Three principal sources of light-induced ROS have been identified:

1. the photosynthetic antennae may transfer excitation energy to molecular oxygen, thereby generating singlet oxygen ($^1O_2$);
2. electrons may leak from reduced ferredoxin in photosystem I (PSI) to molecular oxygen, forming superoxide radicals (O_2^-) via the Mehler reaction; and
3. hydrogen peroxide (H_2O_2) is produced by the peroxysomes during photorespiration.

Primary oxidants rapidly give rise to secondary oxidants. O_2^- disproportionates to H_2O_2, either spontaneously or catalytically by superoxide dismutase. In the presence of transition metals, O_2^- and H_2O_2 together initiate the production of hydroxyl radicals (·OH) through Haber-Weiss reactions (Halliwell and Gutteridge, 1999). Hydroxyl radicals are the most reactive of ROS; they are extremely potent oxidants, and react with organic molecules at near diffusion rates (Mallick and Mohn, 2000). Because of their light-harvesting and oxygen-generating functions and their high concentrations of polyunsaturated fatty acids in thylakoid membranes, chloroplasts under light stress are particularly susceptible to oxidative damage. Any mechanism that reduces the flux of light incident on a chloroplast would reduce the propensity for photooxidative damage.

Possible reductions in ROS production attributable to the light-filtering effects of anthocyanins are difficult to demonstrate *in vivo*. This is because, in addition to their light absorbing properties, anthocyanins also function as cellular antioxidants (see next section), and the relative contributions of each are not easily discriminated. Nonetheless, the effects of light abatement *per se* have been demonstrated recently using isolated chloroplasts suspended in buffered solutions of nitroblue tetrazolium, which yields the blue formazan pigment in proportion to O_2^- concentration (Halliwell and Gutteridge, 1999). Levels of O_2^- in the suspension increased asymptotically over time when the chloroplasts were irradiated with 400 μmol m^{-2} s^{-1} of white light (Neill, 2002). The insertion of a cellulose filter between the light source and the chloroplasts, which reduced the irradiance to 300 μmol m^{-2} s^{-1}, significantly impeded O_2^- generation (Fig. 5A). A red cellulose filter for which the spectral characteristics were similar to those of cyanidin-3-glucoside reduced O_2^- generation more effectively than did a neutral density filter. Chlorophyll bleaching, an early symptom of photooxidative damage, was also significantly retarded (Fig. 5B). O_2^- production was suppressed to an even greater degree when a green cellulose filter was used. However, the absorption of red light by a green filter would also severely compromise photosynthetic performance. The absorption of green quanta by anthocyanins, therefore, provides an efficient mechanism to reduce the generation of O_2^- without impacting appreciably on light capture for photosynthesis.

III. ANTHOCYANINS AS ANTIOXIDANTS

A. MECHANISMS

Anthocyanins are potent antioxidants. Cyanidin, the most prevalent anthocyanidin in leaves (Hrazdina, 1982; Gould and Quinn, 1999) has an antioxidant capacity up to 4.4 times greater than those of ascorbic acid and the vitamin E analogue, Trolox (Rice-Evans *et al.*, 1997). Glycosylation normally diminishes the antioxidant potential, the extent depending on the nature and attachment sites of the sugar(s) (Wang *et al.*, 1997). However, cyanidin-3-glucoside, perhaps the most common foliar anthocyanin, has an antioxidant potency against peroxyl radicals 1.5 times greater than that of cyanidin, and heads the antioxidant league for these pigments (Wang *et al.*, 1997). The high activity has been ascribed primarily to the reducing power of the two ortho-related hydroxyl groups on the B ring of the anthocyanin molecule, the presence of O^+ (oxonium ion) in the C ring, and the numbers and positions of the free hydroxyl groups (van Acker *et al.*, 1996; Rice-Evans *et al.*, 1996). Anthocyanins have been shown to protect against DNA damage and low-

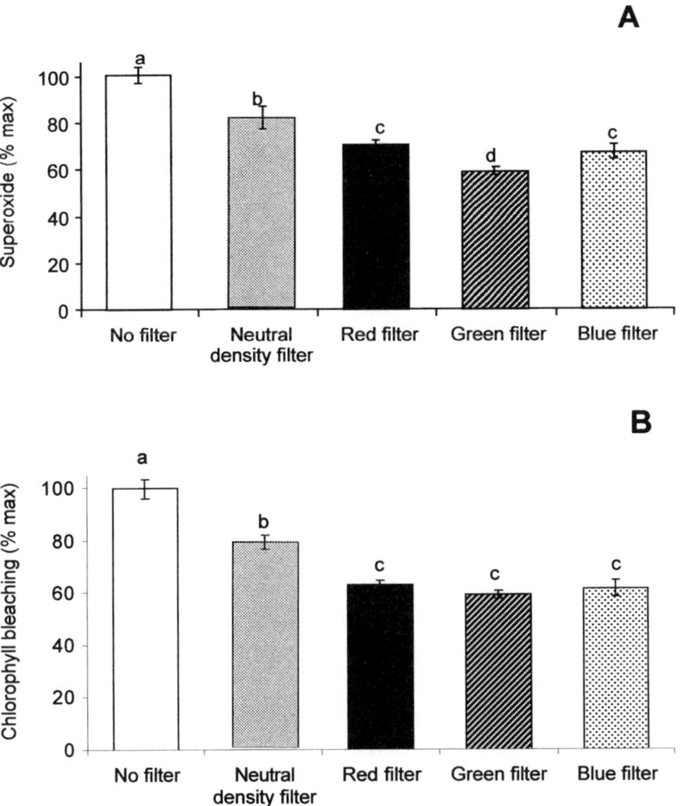

Fig. 5. Superoxide generation (A) and chlorophyll bleaching (B) in chloroplast suspensions from *Lactuca sativa*. Chloroplasts were irradiated for 8 min with 400 μmol m^{-2} s^{-1} white light at 23°C, or else with 300 μmol m^{-2} s^{-1} white, red, green or blue light obtained using cellulose filters. Superoxide levels were estimated by the conversion of nitroblue tetrazolium to formazan, estimated as increments in A_{560} values. Chlorophyll bleaching was quantified as reductions in A_{680} of the chloroplast suspensions. Data (mean ± SE, n = 3) are expressed as percentages of the maximum yield. Values that are significantly different ($P<0.05$) are indicated by different letters. From Neill (2002).

density lipoprotein oxidation (Frankel *et al.*, 1995; Ramirez-Tortosa *et al.*, 2001). Three different anthocyanins isolated from *Phaseolus* seeds each significantly inhibited lipid peroxidation in a rat liposomal system after UV-B irradiation (Tsuda *et al.*, 1996). Thus, anthocyanins have an inherent potential to protect plant cells from the effects of oxidative damage.

Antioxidant capabilities of anthocyanins are achievable through at least two distinct mechanisms:

1. The anthocyanins, like other flavonoids, can chelate to transition metals, particularly the biologically-important ions of copper and iron

(van Acker et al., 1996; Brown *et al.*, 1998; Sarma and Sharma, 1999; Yoshino and Murakami, 1998). By altering the reduction potential of the transition metals, chelation to anthocyanins can decrease the formation of hydroxyl radicals via the Fenton reaction:

$$Fe^{2+} + H_2O_2 \rightarrow \text{intermediate complex} \rightarrow Fe(III) + OH^{\cdot} + OH^{-}$$

Non-chelated transition metals also catalyse the autoxidations of ascorbic acid and a-tocopherol (Halliwell and Gutteridge, 1999). Anthocyanin-transition metal chelates could thus 'spare' these important antioxidants. The ability of anthocyanins to chelate to transition metals has been demonstrated many times *in vitro* (e.g., Bors *et al.*, 1990; George *et al.*, 1999). However, the relative importance of this mechanism *in planta* is unknown.

2. Anthocyanins act directly as antioxidants by virtue of their ability to donate protons. Along with other flavonoids, the anthocyanins can directly scavenge molecular species of active oxygen, including hydrogen peroxide, singlet oxygen, and the superoxide, hydroxyl, and peroxyl radicals (Bors *et al.*, 1994; Yamasaki *et al.*, 1996). They are also effective scavengers of peroxynitrite, a highly reactive oxidant formed when superoxide reacts with nitric oxide (Tsuda *et al.*, 2000). The aryloxyl radicals thus formed may be stabilised through electron delocalisation, or via self-association (Bors *et al.*, 1990; Larson, 1997). Ascorbic acid also has the potential to regenerate the vacuolar anthocyanins by reducing the oxidised product (Yamasaki *et al.*, 1997). As discussed below, recent evidence suggests that the ROS-scavenging abilities of anthocyanins can have biological importance for some red-leafed species.

B. ANTIOXIDANT ACTIVITIES OF RED VERSUS GREEN LEAVES

The abilities of purified solutions of anthocyanins to scavenge ROS *in vitro* do not necessarily translate to an antioxidant advantage of red leaves *in vivo*. Leaves normally hold a suite of enzymatic and low molecular weight antioxidants (LMWAs), any combination of which can serve to protect macromolecules from the effects of ROS (Foyer *et al.*, 1994; Alscher *et al.*, 1997). Environmental conditions that favour anthocyanin formation also stimulate the biosynthesis of other phenolic compounds and enzymic antioxidants such as superoxide dismutase (SOD), catalase (CAT), and ascorbate peroxidase (APX) in leaves (Grace and Logan, 1996; Logan *et al.*, 1998a, b; Sherwin and Farrant, 1998).

Relative contributions of anthocyanins to the antioxidant pool apparently vary across species. For *Elatostema rugosum*, leaf extracts from the red morphs had a demonstrably higher antioxidant status than those from the green morphs (Neill *et al.*, 2002). Methanolic extracts of the red leaves were on average five times more effective at scavenging the α,α-diphenyl-β-picrylhydrazyl (DPPH) radical than the green leaves (Fig. 6A), and anthocyanins contributed to the LMWA pool more than all other constituent phenolics. The red leaves, in addition to their enhanced anthocyanin content, also held the higher levels of SOD, CAT, and APX. In this species, anthocyanins are clearly associated with an enhanced potential to combat the effects of oxidative stress.

In contrast, there were relatively few differences in antioxidant potential between leaves from the red and green morphs of *Quintinia serrata* (Neill, 2002). Both morphs exhibited comparable ranges in LMWA activity against the DPPH radical (Fig. 6B), and they contained similar levels of enzymatic antioxidants. For this species, the league of LMWAs was headed by the hydroxycinnamic acids (predominantly caffeic acid derivatives), followed by the anthocyanins, and then the colourless flavonoids.

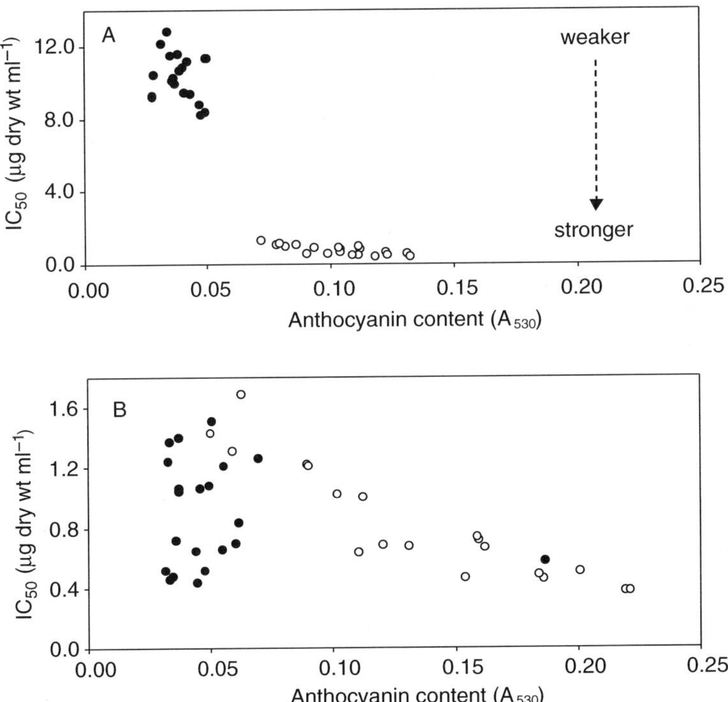

Fig. 6. Antioxidant activities of extracts from red (○) and green (●) leaves of *Elatostema rugosum* (A) and *Quintinia serrata* (B) as a function of anthocyanin content. Activities are shown as concentrations of leaf extract required to scavenge 50% of DPPH radicals (IC_{50} values). The lower IC_{50} values indicate stronger antioxidant activities. Each point represents an individual leaf. Modified from Neill (2002) and Neill *et al.* (2002).

The anthocyanins apparently supplement the antioxidant potential in some *Q. serrata* leaves, but they are not the primary protectants against oxidative damage.

Interspecific differences in the relative contribution of anthocyanins to the antioxidant pool correlate to differences in habitat preference. As a sun-tolerant canopy plant, *Q. serrata* would require protection from New Zealand's harsh UV-B environment. Summertime UV-B irradiance in New Zealand can be up to twice that at similar latitudes in the Northern Hemisphere (Seckmeyer and McKenzie, 1992). Plants can respond rapidly to enhanced UV-B levels by accumulating phenolic protectants (Lois, 1994). The hydroxycinnamic acids are particularly strong absorbers of ultraviolet radiation (Tevini *et al.*, 1991) and their high concentrations in the epidermal cells of *Q. serrata* leaves (Gould *et al.*, 2000) would facilitate both UV-B screening and the scavenging of ROS generated by UV-B exposure. The non-acylated anthocyanins, which reside predominantly in the leaf mesophyll, are relatively weak attenuators of UV-B, but would serve to scavenge additional ROS generated by the chloroplasts. By contrast, *E. rugosum* is normally a shade-adapted plant with a comparatively low requirement for UV-B protection. Shade plants are, however, especially vulnerable to photoinhibitory and photooxidative effects of transient sunflecks (Le Gouallec *et al.*, 1990). The complement of five different anthocyanins in the mesophyll of leaves of *E. rugosum* (Neill *et al.*, 2002) would confer robust protection from photooxidation and other oxidative stressors.

C. H_2O_2 SCAVENGING *IN VIVO*

If the antioxidant capabilities apparent in purified extracts are also realised *in vivo*, then anthocyanins would provide a formidable line of defence against oxidative stress for shoots under unfavourable environments. It has been argued, however, that anthocyanins are not optimally located within the plant cell for ROS scavenging (Yamasaki *et al.*, 1997). Anthocyanins are synthesised in the cytoplasm, but they are rapidly transported across the tonoplast into the cell vacuole (Marrs *et al.*, 1995), where they are physically isolated from cytoplasmic sources of active oxygen. Most oxygen radicals cannot readily permeate the tonoplast (Takahashi and Asada, 1983), although in aqueous solution superoxide is rapidly protonated to the hydroperoxyl radical, or else converted by SOD to H_2O_2, both of which can freely enter the cell vacuole (Yamasaki *et al.*, 1997). However, it may be that cytosolic and organelle-bound antioxidants, rather than the vacuolar anthocyanins, offer the first line of defence against oxidative stress.

Confirmation of an active role for anthocyanins as antioxidants in leaves requires the demonstration of ROS scavenging by cyanic cells *in situ*. This has recently been achieved in part using the leaves of *Pseudowintera colorata* (Winteraceae), an erect shrub from New Zealand. The species typically holds anthocyanins along the leaf margins, and in patches irregularly distributed across the adaxial surface of the lamina (Plate 8 C). Oxidative responses of the red and green portions of *P. colorata* leaf laminae were compared by the real-time imaging of H_2O_2 in cells after mechanical injury (Gould *et al.*, 2002b). Paradermal sections through the laminae, each containing the adaxial epidermis and uppermost palisade mesophyll, were punctured using a fine needle, and examined by fluorescence microscopy. Changes in levels of cellular H_2O_2 were monitored using two fluorochromes, dichlorofluorescein and scopoletin.

Mechanical injury to *P. colorata* leaf laminae elicited an oxidative burst almost immediately from chloroplasts in the palisade mesophyll. A band of H_2O_2 rapidly enveloped the wound and diffused centrifugally across a radius of approximately 20 cells. The H_2O_2, rendered visible by the emission of green fluorescence (dichlorofluorescein) or by the extinction of blue fluorescence (scopoletin), evolved from both green and red regions of the punctured laminae. However, differences between the two regions became apparent within minutes of injury (Fig. 7). H_2O_2 continued to accumulate in the chlorenchyma of green lamina regions for a further 10 min, and then decreased only slowly. By contrast, levels of

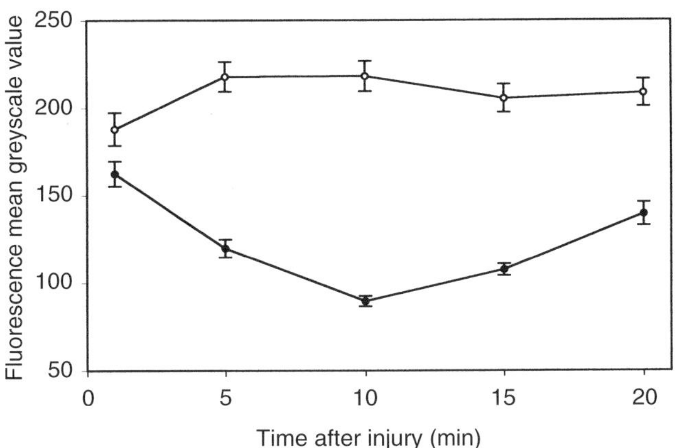

Fig. 7. Fluorescence intensities of scopoletin-infused paradermal sections from leaves of *Pseudowintera colorata* after mechanical injury. Fluorescence was quantified as mean greyscale values (1 to 255) from digitally-captured images of red (○) and green (●) portions of laminae. The higher fluorescence levels are indicative of lower H_2O_2 concentrations. n = 3 for each treatment. Bars show standard errors. Modified from Gould *et al.* (2002b).

H_2O_2 in the red regions declined rapidly to background counts within the first 5 min, and consistently low levels were maintained thereafter. Although the red regions were enriched in anthocyanins, flavonols, dihydroflavonols, and hydroxycinnamic acids relative to the green regions, only the anthocyanins had suitable cellular locations (in the palisade mesophyll) to account for the enhanced rates of H_2O_2 scavenging. The depletion of H_2O_2 from green regions could be accelerated to rates approaching those of red regions by infusing the cells with *N*-acetyl-L-cysteine, a general scavenger of ROS. The data therefore support the hypothesis that red cells have elevated antioxidant capabilities *in planta*.

The intracellular site at which anthocyanins operate as antioxidants has yet to be determined. H_2O_2 might be scavenged by the cytosolic, colourless equilibrium of anthocyanin tautomers prior to their transport into the vacuole, or else by the more abundant, pigmented forms of anthocyanins in the vacuolar solution. Purified anthocyanin fractions exhibit strong antioxidant activities both at acidic pH values characteristic of the vacuole, and at the cytosolic pH 7 (Lapidot *et al.*, 1999; Neill *et al.*, 2002). Yamasaki (1997) argued that the vacuolar anthocyanins would assume increasing importance under conditions of severe stress or during periods of rapid plant growth, as H_2O_2 leaks from cellular organelles and diffuses across the tonoplast membrane. Vacuoles isolated from mesophyll protoplasts of *Vicia faba* have a demonstrable capacity to reduce H_2O_2 (Takahama and Egashira, 1990). This has been ascribed to the peroxidase-mediated scavenging of H_2O_2 by flavonoids and hydroxycinnamic acids located in the cell vacuole (Takahama and Oniki, 1997; Yamasaki, 1997;Yamasaki *et al.*, 1997).

IV. ANTHOCYANINS AS MEDIATORS OF PLANT STRESS RESPONSES

Oxidative stress is not restricted to conditions of high light or mechanical injury. A diverse assortment of environmental stimuli, both biotic and abiotic, has been found to elicit ROS formation in plants. These include: drought stress (Smirnoff, 1993; Sgherri and Navari-Izzo, 1995; Loggini *et al.*, 1999); extremes of temperature (Doke *et al.*, 1994; Rao *et al.*, 1995); heavy metal toxicity (Weckx and Clijsters, 1996; Sandalio *et al.*, 2001); salinity (Meneguzzo *et al.*, 1999); UV radiation (Murphy and Huerta, 1990); exposure to ozone (Melhorn, 1990; Sharma *et al.*, 1996); herbicide treatments (Tanaka, 1994) and pathogenic attack (Low and Merida, 1996).

All of these stressors have also been reported to induce anthocyanins in plants! Anthocyanins have long been considered markers of generalised stress responses in leaves and other vegetative organs (McClure,

1975; Hrazdina, 1982). The transcription and expression of genes involved in anthocyanin biosynthesis are upregulated by stressors as diverse as low temperatures (Christie *et al.*, 1994; Hasegawa *et al.*, 2001), water stress (Gopalakrishna *et al.*, 2001), high cadmium concentrations (Marrs and Walbot, 1997), fungal elicitors (Schmid *et al.*, 1990), and strong irradiance (Jackson *et al.*, 1995; Leyva *et al.*, 1995). Moreover, an enhanced tolerance of adverse environments is often associated with, although not necessarily ascribed to, anthocyanin formation in vegetative organs (Chalker-Scott, 1999).

It is possible, therefore, that anthocyanins serve to ameliorate the defence mechanisms in plants that are subjected to some, or all, of these environmental inducers of oxidative stress. Accordingly, the production of anthocyanins would fit neatly into Leshem and Kuiper's (1996) definition of a *general adaptation syndrome*, for which different types of stress evoke similar adaptation responses. Along with compounds such as abscisic acid, jasmonates, ethylene, and heat shock proteins, the anthocyanins may function as general mitigators of oxidative damage. We should add the caveat, however, that as yet there is no direct evidence that plants benefit from the antioxidant properties of anthocyanins. Clearly, anthocyanins have the potential to scavenge ROS, and they have been shown many times to reduce lipid peroxidation in a test tube – but we lack solid, experimental data on the impact of anthocyanins on the health of plant organelles *in vivo*. Future research should include *in vivo* appraisals of lipid membrane integrity in cyanic cells under natural environmental conditions.

V. OVERVIEW AND CONCLUDING REMARKS

Anthocyanins have the potential to reduce oxidative stress in leaves via three mechanisms. First, they can directly scavenge most species of reactive oxygen, often more effectively than the other flavonoids. Second, the anthocyanins (and other flavonoids) can chelate to transition metals, thereby preventing the autoxidation of other key antioxidants, and reducing the formation of superoxide through the Fenton reaction. Third, by absorbing high-energy quanta that would otherwise strike the chloroplasts, the anthocyanins would reduce the formation of $^1O_2$, O_2^- and H_2O_2 via photooxidative and photorespiratory processes.

The antioxidant hypothesis for anthocyanins is attractive because it potentially explains at least some of the reported variation in location and timing of red pigmentation in leaves, as well as the diversity in environmental triggers. Irregular red blotches on the lamina surface, for example, might be attributable to localised oxidative stress resulting from fungal infection or herbivory (Costa-Arbulú *et al.*, 2001; Stone *et al.*,

2001), and anthocyanins on leaf undersurfaces of tropical understorey plants would be suitably located to provide antioxidant protection without compromising light capture. Anthocyanins in the young, flushing leaves of tropical canopy trees might protect developing chloroplasts against photooxidation until organelle-bound antioxidants are formed (Polle, 1997). Similarly, the optical masking of chlorophyll by anthocyanins in senescing leaves of deciduous trees might increase the efficiency of nutrient retrieval by reducing the risk of photooxidative damage (Feild *et al.*, 2001).

Can antioxidant protection be considered a unifying explanation for anthocyanins in leaves? Clearly, the antioxidant capabilities of anthocyanins are important to some species (e.g. *Elatostema rugosum*) more than others (e.g. *Quintinia serrata*), but at present, there are insufficient data on the antioxidant activities of cyanic leaves from disparate taxonomic groups to judge the wholesale applicability of the hypothesis. It is also noteworthy that the anthocyanins are a particularly ancient group of flavonoid pigments. Anthocyanins and the structurally-related 3-deoxyanthocyanins are present in some ferns and mosses (Markham, 1988; Cooper-Driver and Bhattacharya, 1998), and have even been reported in liverworts (Post and Vesk, 1992; Kunz *et al.*, 1994), the most primitive land plants (Qui *et al.*, 1998). It seems highly likely, therefore, that over the course of about 470 million years of evolution these compounds have developed various functions that contribute in different ways and to different degrees to the physiology of plants. Thus, anthocyanins in arctic plants could serve both to elevate leaf temperature as well as to reduce oxidative damage. Similarly, anthocyanins in the brown leaves of some seedlings could function as agents of camouflage, exert anti-fungal properties, and also scavenge free radicals generated by herbivory. All of the putative protective functions of anthocyanins – including the amelioration of defence against oxidative stress – are likely to be crucial to the survival of a species under their unique sets of environmental conditions. Anthocyanins are multi-functional. There is probably no unified explanation for anthocyanins in leaves.

ACKNOWLEDGEMENTS

Studies on leaf optics and antioxidant properties in our laboratories were supported by a Marsden grant (UOA 707) from the Royal Society of New Zealand, and the National Science Foundation (NSF DBI-9724499). We are particularly grateful to Dr. Kenneth Markham (Industrial Research Ltd, Lower Hutt, New Zealand) for his advice and constructive comments on this manuscript.

REFERENCES

van Acker, S. A. B. E, van den Berg, D.-J., Tromp, M. N. J. L., Griffioen, D. H., van Bennekom, W. P., van der Vugh, W. J. F. and Bast, A. (1996). Structural aspects of antioxidant activity in flavonoids. *Free Radical Biology and Medicine* **20**, 331–342.

Alscher, R. G., Donahue, J. L. and Cramer, C. L. (1997). Reactive oxygen species and antioxidants: relationships in green cells. *Physiologia Plantarum* **100**, 224–233.

Barker, D. H., Seaton, G. G. R. and Robinson, S. A. (1997). Internal and external photoprotection in developing leaves of the CAM plant *Cotyledon orbiculata. Plant, Cell and Environment* **20**, 617–624.

Björkman, O. (1981). Responses to different quantum flux densities. *In*: "Encyclopedia of Plant Physiology, N. S., Vol. 12A" (O. L. Lange, P. S. Nobel, C. B. Osmond and H. Ziegler, eds) pp. 57–107. Springer-Verlag, Berlin.

Bors, W., Heller, W., Michel, C. and Saran, M. (1990). Flavonoids as antioxidants: determination of radical scavenging efficiencies. *Methods in Enzymology* **186**, 343–355.

Bors W., Michel, C. and Saran, M. (1994). Flavonoid antioxidants: rate constants for reactions with oxygen radicals. *Methods in Enzymology* **234**, 420–429.

Brouillard, R. and Dangles, O. (1994). Flavonoids and flower colour. *In* "The Flavonoids – Advances in Research Since 1986" (J. B. Harborne, ed.) pp. 565–586. Chapman and Hall, London.

Brown, J. E., Khodr, H., Hider, R. C. and Rice-Evans, C. A. (1998). Structural dependence of flavonoid interactions with Cu^{2+} ions: implications for their antioxidant properties. *Biochemical Journal* **330**, 1173–1178.

Burger, J. and Edwards, G. E. (1996). Photosynthetic efficiency and photodamage by UV and visible radiation, in red versus green leaf *Coleus* varieties. *Plant Cell Physiology* **73**, 395–399.

Chalker-Scott, L. (1999). Environmental significance of anthocyanins in plant stress responses. *Photochemistry and Photobiology* **70**, 1–9.

Cherepy, N. J., Smestad, G. P., Grätzel, M. and Zhang, J. Z. (1997). Ultrafast electron injection: implications for a photoelectrochemical cell utilizing an anthocyanin dye-sensitized TiO_2 nanocrystalline electrode. *Journal of Physical Chemistry B* **101**, 9342–9351.

Choinski Jr, J. S. and Johnson, J. M. (1993). Changes in photosynthesis and water status of developing leaves of *Brachystegia spiciformis* Benth. *Tree Physiology* **13**, 17–27.

Christie, P. J., Alfenito, M. R. and Walbot, V. (1994). Impact of low-temperature stress on general phenylpropanoid and anthocyanin pathways: enhancement of transcript abundance and anthocyanin pigmentation in maize seedlings. *Planta* **194**, 541–549.

Cooper-Driver, G. A. and Bhattacharya, M. (1998). Role of phenolics in plant evolution. *Phytochemistry* **49**, 1165–1174.

Costa-Arbulú, C., Gianoli, E., Gonzáles, W. and Niemeyer, H. M. (2001). Feeding by the aphid *Sipha flava* produces a reddish spot on leaves of *Sorghum halepense*: an induced defence? *Journal of Chemical Ecology* **27**, 273–283.

Dhawale, N. M., Akhtar, M. and Sharma, V. (1983). Hill reaction activity of chloroplasts from *Amaryllis vittata* flowers and its enhancement by anthocyanin supplementation. *Photosynthetica* **17**, 264–266.

Dodd, I. C., Critchley, C., Woodall, G. S. and Stewart, G. R. (1998). Photoinhibition in differently coloured juvenile leaves of *Syzygium* species. *Journal of Experimental Botany* **49**, 1437–1445.

Doke, N., Miura, Y., Leandro, M. S. and Kawakita, K. (1994). Involvement of superoxide in signal transduction: responses to attack by pathogens, physical and chemical shocks and UV-irradiation. *In* "Causes of Photooxidative Stress and Amelioration of Defense Systems in Plants" (C. H. Foyer and P. Mullineaux, eds) pp. 177–197. CRC Press Inc., Boca Raton.

Drabent, R., Pliszka, B. and Olszewska, T. (1999). Fluorescence properties of plant anthocyanin pigments. I. Fluorescence of anthocyanins in *Brassica oleracea* L. extracts. *Journal of Photochemistry and Photobiology B: Biology* **50**, 53–58.

Eller, B. M., Glättli, R. and Flach, B. (1981). Optical properties and pigments of sun and shade leaves of the beech (*Fagus silvatica* L.) and the copper-beech (*Fagus silvatica* cv. Atropunicea). *Flora (Jena)* **171**, 170–185.

Feild, T. S., Lee, D. W. and Holbrook, N. M. (2001). Why leaves turn red in autumn. The role of anthocyanins in senescing leaves of red-osier dogwood. *Plant Physiology* **127**, 566–574.

Foyer, C. H., Lelandais, M. and Kunert, K. H. (1994). Photooxidative stress in plants. *Physiologia Plantarum* **92**, 696–717.

Frankel, E. F., Waterhouse, A. L. and Teissedre, P. L. (1995). Principal phenolic phytochemicals in selected California wines and their antioxidant activity in inhibiting oxidation of human low-density lipoproteins. *Journal of Agricultural Food Chemistry* **43**, 890–894.

Gausman, H. W. (1982). Visible light reflectance, transmittance, and absorptance of differently pigmented cotton leaves. *Remote Sensing of Environment* **13**, 233–238.

George, F., Figueiredo, P. and Brouillard, R. (1999). Malvin Z-chalcone: an unexpected new open cavity for the ferric cation. *Phytochemistry* **50**, 1391–1394.

Gopalakrishna, R., Kumar, G., KrishnaPrasad, B. T., Mathew, M. K. and Kumar, M. U. (2001). A stress-responsive gene from groundnut, Gdi-15, is homologous to flavonol 3-O-glucosyltransferase involved in anthocyanin biosynthesis. *Biochemical and Biophysical Research Communications* **284**, 574–579.

Gould, K. S. and Quinn, B. D. (1999). Do anthocyanins protect leaves of New Zealand native species from UV-B? *New Zealand Journal of Botany* **37**, 175–178.

Gould, K. S., Kuhn, D. N., Lee, D. W. and Oberbauer, S. F. (1995). Why leaves are sometimes red. *Nature* **378**, 241–242.

Gould, K. S., Markham, K. R., Smith, R. H. and Goris, J. J. (2000). Functional role of anthocyanins in the leaves of *Quintinia serrata* A. Cunn. *Journal of Experimental Botany* **51**, 1107–1115.

Gould, K. S., Vogelmann, T. C., Han, T. and Clearwater, M. J. (2002a). Profiles of photosynthesis within red and green leaves of *Quintinia serrata* A. Cunn. *Physiologia Plantarum* **116**, 127–133.

Gould, K. S., McKelvie, J. and Markham, K. R. (2002b). Do anthocyanins function as antioxidants in leaves? Imaging of H_2O_2 in red and green leaves after mechanical injury. *Plant, Cell and Environment* **25**, 1261–1269.

Grace, S. C. and Logan, B. A. (1996). Acclimation of foliar antioxidant systems to growth irradiance in three broad-leaved evergreen species. *Plant Physiology* **112**, 1631–1640.

Hackett, W. P. (1985). Juvenility, maturation and rejuvenation in woody plants. *Horticultural Reviews* **7**, 109–155.

Halliwell, B. and Gutteridge, J. M. C. (1999). "Free Radicals in Biology and Medicine", 3rd edition, pp. 936. Oxford University Press, Oxford.

Han, T. and Vogelmann, T. C. (1999). A photoacoustic spectrometer for measuring heat dissipation and oxygen quantum yield at the microscopic level within leaf tissues. *Journal of Photochemistry and Photobiology B: Biology* **48**, 158–165.

Harborne, J. B. (1967). The anthocyanin pigments. *In* "Comparative Biochemistry of the Flavonoids" (J. B. Harborne, ed.), pp. 1–36. Academic Press, London.

Hasegawa, H., Fukasawa-Akada, T., Okuno, T., Niizeki, M. and Suzuki, M. (2001). Anthocyanin accumulation and related gene expression in Japanese parsley (*Oenanthe stolonifera*, DC.) induced by low temperature. *Journal of Plant Physiology* **158**, 71–78.

Hinchliff, T. (2001). Seasonal variability in anthocyanin concentration in relation to photosynthetic parameters and antioxidant ability in native New Zealand species. M.Sc. thesis, University of Auckland, New Zealand.

Hoch, W. A., Zeldin, E. L. and McCown, B. H. (2001). Physiological significance of anthocyanins during autumnal leaf senescence. *Tree Physiology* **21**, 1–8.

Hrazdina, G. (1982). Anthocyanins. *In* "The Flavonoids: Advances in Research" (J. B. Harborne and T. J. Mabry, eds), pp. 135–188. Chapman and Hall, London.

Jackson, J. A., Fuglevand, G., Brown, B. A., Shaw, M. J. and Jenkins, G. I. (1995). Isolation of *Arabidopsis* mutants altered in the light-regulation of chalcone synthase gene expression using a transgenic screening approach. *Plant Journal* **8**, 369–380.

Krol, M., Gray, G. R., Hurry, V. M., Oquist, G., Malek, L. and Huner, N. P. (1995). Low-temperature stress and photoperiod affect an increased tolerance to photoinhibition in *Pinus banksiana* seedlings. *Canadian Journal of Botany* **73**, 1119–1127.

Kunz, S., Burkhardt, G. and Becker, H. (1994). Riccionidins A and B, anthocyanidins from the cell walls of the liverwort *Ricciocarpos natans*. *Phytochemistry* **35**, 233–235.

Lapidot, T., Harel, S., Akiri, B., Granit, R. and Kanner, J. (1999). pH-dependent forms of red wine anthocyanins as antioxidants. *Journal of Agricultural and Food Chemistry* **47**, 67–70.

Larson, R. (1997). "Naturally Occurring Antioxidants", pp. 195. Lewis Publisher, Boca Raton.

Lee, D. W. and Collins, T. W. (2001). Phylogenetic and ontogenetic influences on the distribution of anthocyanins and betacyanins in leaves of tropical plants. *International Journal of Plant Science* **162**, 1141–1153.

Lee, D. W. and Lowry, J. B. (1980). Young-leaf anthocyanin and solar ultraviolet. *Biotropica* **12**, 75–76.

Lee, D. W., Lowry, J. B. and Stone, B. C. (1979). Abaxial anthocyanin layer in leaves of tropical rain forest plants: enhancer of light capture in deep shade. *Biotropica* **11**, 70–77.

Lee, D. W., Brammeier, S. and Smith, A. P. (1987). The selective advantages of anthocyanins in developing leaves of mango and cacao. *Biotropica* **19**, 40–49.

Le Gouallec, J. L., Cornic, G. and Blanc, P. (1990). Relations between sunfleck sequences and photoinhibition of photosynthesis in a tropical rain forest understory herb. *American Journal of Botany* **77**, 999–1006.

Leshem, Y. Y. and Kuiper, P. J. C. (1996). Is there a GAS (general adaptation syndrome) response to various types of environmental stress? *Biologia Plantarum* **38**, 1–18.

Leyva, A., Jarillo, J. A., Salinas, J. and Maartinez-Zapater, J. M. (1995). Low temperature induces the accumulation of phenylanaline ammonia-lyase and chalcone synthase mRNAs of *Arabidopsis thaliana* in a light-dependent manner. *Plant Physiology* **108**, 39–46.

Logan, B. A., Demmig-Adams, B., Adams Iii, W. W. and Grace, S. (1998a). Antioxidants and xanthophyll cycle-dependant energy dissipation in *Cucurbita pepo* L. and *Vinca major* L. acclimated to four growth PPFDs in the field. *Journal of Experimental Botany* **49**, 1869–1879.

Logan, B. A., Grace, S. C., Adams Iii, W. W. and Demmig-Adams, B. (1998b). Seasonal differences in xanthophyll cycle characteristics and antioxidant in *Mahonia repens* growing in different light environments. *Oecologia* **116**, 9–17.

Loggini, B., Scartazza, A., Brugnoli, E. and Navari-Izzo, F. (1999). Antioxidative defense system, pigment composition and photosynthetic efficiency in two wheat cultivars subjected to drought. *Plant Physiology* **199**, 1091–1099.

Lois, R. (1994). Accumulation of UV-absorbing flavonoids induced by UV-B radiation in *Arabidopsis thaliana* L. I. Mechanisms of UV-resistance in *Arabidopsis*. *Planta* **194**, 498–503.

Low, P. S. and Merida, J. R. (1996). The oxidative burst in plant defense: function and signal transduction. *Physiologia Plantarum* **96**, 533–542.

Mallick, N. and Mohn, F. H. (2000). Reactive oxygen species: response of algal cells. *Journal of Plant Physiology* **157**, 183–193.

Markham, K. R. (1988). Distribution of flavonoids in the lower plants and its evolutionary significance. *In* "The Flavonoids" (J. B. Harborne, ed.), pp. 427–468. Chapman and Hall, London.

Markham, K. R., Gould, K. S., Winefield, C. S., Mitchell, K. A., Bloor, S. J. and Boase, M. R. (2000). Anthocyanic vacuolar inclusions – their nature and significance in flower colouration. *Phytochemistry* **55**, 327–336.

Marrs, K. A. and Walbot, V. (1997). Expression and RNA splicing of the maize glutathione S-transferase Bronze2 gene is regulated by cadmium and other stresses. *Plant Physiology* **113**, 93–102.

Marrs, K. A., Alfenito, R. R., Lloyd, A. M. and Walbot, V. (1995). A glutathione S-transferase involved in vacuolar transfer encoded by the maize gene *Bronze-2*. *Nature* **375**, 397–400.

McClure, J. W. (1975). Physiology and function of flavonoids. *In* "The Flavonoids" (J. B. Harborne, T. J. Mabry and H. Mabry, eds), pp. 970–1055. Chapman and Hall, London.

Melhorn, H. (1990). Ethylene-promoted ascorbate peroxidase activity protects plants against hydrogen peroxide, ozone and paraquat. *Plant, Cell and Environment* **13**, 971–976.

Meneguzzo, S., Navari-Izzo, F. and Izzo, R. (1999). Antioxidative responses of shoots and roots of wheat to increasing NaCl concentrations. *Journal of Plant Physiology* **155**, 274–280.

Murphy, T. M. and Huerta, A. J. (1990). Hydrogen peroxide formation in cultured rose cells in response to UV-C radiation. *Physiologia Plantarum* **78**, 247–253.

Neill, S. O. (2002). The functional role of anthocyanins in leaves. Ph.D. thesis, University of Auckland, New Zealand.

Neill, S. O. and Gould, K. S. (1999). Optical properties of leaves in relation to anthocyanin concentration and distribution. *Canadian Journal of Botany* **77**, 1777–1782.

Neill, S. O., Gould, K. S., Kilmartin, P. A., Mitchell, K. A. and Markham, K. R. (2002). Antioxidant activities of red versus green leaves in *Elatostema rugosum*. *Plant, Cell and Environment* **25**, 539–547.

Nishio, J. N. (2000). Why are higher plants green? Evolution of the higher plant photosynthetic pigment complement. *Plant, Cell and Environment* **23**, 539–548.

Nozue, M., Kubo, H., Nishimura, T. and Yasuda, H. (1995). Detection and characterisation of a vacuolar protein (VP24) in anthocyanin-producing cells of sweet potato in suspension culture. *Plant and Cell Physiology* **36**, 883–889.

Peckett, R. C. and Small, C. J. (1980). Occurrence, location and development of anthocyanoplasts. *Phytochemistry* **19**, 2571–2576.

Polle, A. (1997). Defense against photooxidative damage in plants. *In* "Oxidative Stress and the Molecular Biology of Antioxidant Defenses" (J. G. Scandalios, ed.), pp. 623–666. Cold Spring Harbor Laboratory Press, New York.

Post, A. and Vesk, M. (1992). Photosynthesis, pigments, and chloroplast ultrastructure of an Antarctic liverwort from sun-exposed and shaded sites. *Canadian Journal of Botany* **70**, 2259–2264.

Qiu, Y.-L., Cho, Y., Cox, J. C. and Palmer, J. D. (1998). The gain of three mitochondrial introns identifies liverworts as the earliest land plants. *Nature* **394**, 671–674.

Ramirez-Tortosa, C., Andersen, O. M., Gardner, P. T., Morrice, P. C., Wood, S. G., Duthie, S. J., Collins, A. R. and Duthie, G. G. (2001). Anthocyanin-rich extract decreases indices of lipid peroxidation and DNA damage in vitamin E-depleted rats. *Free Radical Biology and Medicine* **31**, 1033–1037.

Rao, M. V., Halle, B. A. and Ormrod, D. P. (1995). Amelioration of ozone-induced oxidative damage in wheat plants grown under high carbon dioxide. *Plant Physiology* **109**, 421–432.

Rice-Evans, C. A., Miller, N. J. and Paganga, G. (1996). Structure-antioxidant activity relationships of flavonoids and phenolic acids. *Free Radical Biology and Medicine* **20**, 933–956.

Rice-Evans, C. A., Miller, N. J. and Paganga, G. (1997). Antioxidant properties of phenolic compounds. *Trends in Plant Science* **2**, 152–159.

Sandalio, L. M., Dalurzo, H. C., Gomez, M., Romero-Puertas, M. C. and del Rio, L. A. (2001). Cadmium-induced changes in the growth and oxidative metabolism of pea plants. *Journal of Experimental Botany* **52**, 2115–2126.

Sarma, A. D. and Sharma, R. (1999). Anthocyanin-DNA copigmentation complex: mutual protection against oxidative damage. *Phytochemistry* **52**, 1313–1318.

Schmid, J., Doerner, P. W., Clouse, S. D., Dixon, R. A. and Lamb, C. J. (1990). Developmental and environmental regulation of a bean chalcone synthase promoter in transgenic tobacco. *Plant Cell* **2**, 619–632.

Seckmeyer, G. and McKenzie, R. L. (1992). Increased ultraviolet radiation in New Zealand (45°S) relative to Germany (48°N). *Nature* **359**, 135–137.

Sgherri, C. L. and Navari-Izzo, F. (1995). Sunflower seedlings subjected to increasing water deficit stress: oxidative stress and defense mechanisms. *Physiologia Plantarum* **93**, 25–30.

Sharma, V. and Banerji, D. (1981). Enhancement of Hill activity by anthocyanins under both 'white incandescent' and green irradiation. *Photosynthetica* **15**, 540–542.

Sharma, Y. K., Leon, J., Raskin, I. and Davis, K. R. (1996). Ozone-induced responses in *Arabidopsis thaliana*: the role of salicylic acid in the accumulation of defense-related transcripts and induced resistance. *Proceedings of the National Academy of Science, USA* **93**, 5099–5104.

Sherwin, H. W. and Farrant, J. M. (1998). Protection mechanisms against excess light in the resurrection plants *Craterostigma wilmsii* and *Xerophyta viscosa. Plant Growth Regulation* **24**, 203–210.

Smillie, R. M. and Hetherington, S. E. (1999). Photoabatement by anthocyanin shields photosynthetic systems from light stress. *Photosynthetica* **36**, 451–463.

Smirnoff, N. (1993). The role of active oxygen in the response of plants to water deficit and desiccation. *New Phytologist* **125**, 27–58.

Smith, R. H. (2000). Effects of anthocyanin pigments on light capture and photosynthesis in *Elatostema rugosum* A. Cunn. leaves. M.Sc. thesis, University of Auckland, New Zealand.

Smith, W. K., Vogelmann, T. C., DeLucia, E. H., Bell, D. T. and Shepherd, K. A. (1997). Leaf form and photosynthesis: do leaf structure and orientation interact to regulate internal light and carbon dioxide? *BioScience* **47**, 785–793.

Stone, C., Chisholm, L. and Coops, N. (2001). Spectral reflectance characteristics of eucalypt foliage damaged by insects. *Australian Journal of Botany* **49**, 687–698.

Sun, J., Nishio, J. N. and Vogelmann, T. C. (1998). Green light drives CO_2 fixation deep within leaves. *Plant, Cell and Environment* **39**, 1020–1026.

Takahama, U. and Egashira, T. (1990). Hydrogen peroxide-dependent oxidation of 3,4-dihydroxyphenylalanine in vacuoles of mesophyll cells of *Vicia faba* L. Participation of peroxidase in the oxidation. *Plant Cell Physiology* **31**, 539–544.

Takahama, U. and Oniki, T. (1997). A peroxidase/phenolics/ascorbate system can scavenge hydrogen peroxide in plant cells. *Physiologia Plantarum* **101**, 845–852.

Takahashi, M. A. and Asada, K. (1983). Superoxide anion permeability of phospholipid membranes and chloroplast thylakoids. *Archives of Biochemistry and Biophysics* **226**, 558–566.

Tanaka, K. (1994). Tolerance to herbicides and air pollutants. *In* "Causes of Photooxidative Stress and Amelioration of Defense Systems in Plants" (C. H. Foyer and P. Mullineaux, eds), pp. 365–378. CRC Press Inc., Boca Raton.

Tevini, M., Braun, J. and Fieser, G. (1991). The protective function of the epidermal layer of rye seedlings against ultraviolet-B radiation. *Photochemistry and Photobiology* **53**, 329–333.

Tsuda, T., Shiga, K., Ohshima, K., Kawakishi, S. and Osawa, T. (1996). Inhibition of lipid peroxidation and the active oxygen radical scavenging effect of anthocyanin pigments isolated from *Phaseolus vulgaris* L. *Biochemical Pharmacology* **52**, 1033–1039.

Tsuda, T., Kato, Y. and Osawa, T. (2000). Mechanism for the peroxynitrite scavenging activity by anthocyanins. *FEBS Letters* **484**, 207–210.

Tuohy, J. M. and Choinski Jr, J. S. (1990). Comparative photosynthesis in developing leaves of *Brachystegia spiciformis* Benth. *Journal of Experimental Botany* **41**, 919–923.

Vogelmann, T. C. and Han, T. (2000). Measurement of gradients of absorbed light in spinach leaves from chlorophyll fluorescence profiles. *Plant, Cell and Environment* **23**, 1303–1311.

Wang, H., Cao, G. and Prior, R. L. (1997). Oxygen radical absorbing capacity of anthocyanins. *Journal of Agricultural Food Chemistry* **45**, 304–309.

Weckx, J. E. J. and Clijsters, H. (1996). Oxidative damage and defense mechanisms in primary leaves of *Phaseolus vulgaris* as a result of root assimilation of toxic amounts of copper. *Physiologia Plantarum* **96**, 506–512.

Woodall, G. S., Dodd, I. C. and Stewart, G. R. (1998). Contrasting leaf development within the genus *Syzygium*. *Journal of Experimental Botany* **49**, 79–87.

Yamasaki, H. (1997). A function of colour. *Trends in Plant Science* **2**, 7–8.

Yamasaki, H., Uefuji, H. and Sakihama, Y. (1996). Bleaching of the red anthocyanin induced by superoxide radical. *Archives of Biochemistry and Biophysics* **332**, 183–186.

Yamasaki, H., Sakihama, Y. and Ikehara, N. (1997). Flavonoid-peroxidase reaction as a detoxification mechanism of plant cells against H_2O_2. *Plant Physiology* **115**, 1405–1412.

Yoshino, M. and Murakami, K. (1998). Interaction of iron with polyphenolic compounds: application to antioxidant characterization. *Analytical Biochemistry* **257**, 40–44.

AUTHOR INDEX

Numbers in **bold** refer to pages on which full references are listed

C

T

SUBJECT INDEX

D

E

F

M

N

O

P

W

X

Z